THE MEANING OF LIFE

THE MEANING
OF LIFE

Edited by

E. D. KLEMKE

Professor of Philosophy
Iowa State University

New York   Oxford
OXFORD UNIVERSITY PRESS
1981

Library of Congress Cataloging in Publication Data
Main entry under title:
The Meaning of life.
Bibliography: p.
Includes index.
1. Life—Addresses, essays, lectures.
I. Klemke, E. D., 1926–
BD431.M4688   128   80–20132   ISBN 0–19–502871–6

Printed in the United States of America

To Michael D. Hogue

Preface

Viktor Frankl once wrote: "Man's concern about a meaning of life is the truest expression of the state of being human."

The selections contained in this volume are, I believe, the most important essays (or chapters) that have been written on the topic of the meaning of life. All but one of the selections are by twentieth-century writers, philosophers, theologians, and others. In some collections of essays, the editor provides a brief summary of all of the selections contained in the anthology. I shall not attempt to do that here, for I believe that the readers' delight and stimulation will be enhanced if they turn to the selections and let the authors speak for themselves.

I would like to express my gratitude to all those who have helped with regard to the planning and preparation of this book. Among those to whom I am especially grateful are: John Elrod, Paul Taylor, John McCall, Ted Solomon, Steven Isaacson, Rowena Wright, Bernice Power, and Annette Van Cleave. Finally, to the members of Alpha Iota may I say: Thank you for your constant kindness.

Ames, Iowa   E. D. K.
July, 1980

Acknowledgments

The editor gratefully acknowledges the kind permission of the authors, editors, and publishers that have enabled him to print the essays included in this book.

PART I

From Leo Tolstoy, *My Confession.* Translated by Leo Wiener. London: J. M. Dent and Sons, 1905. Reprinted by permission of the publisher.

David F. Swenson, "The Dignity of Human Life." From *Kierkegaardian Philosophy in the Faith of a Scholar.* Philadelphia: Westminster Press, 1949. Pp. 13–28.

David F. Swenson, "The Transforming Power of Otherworldliness." From *Kierkegaardian Philosophy in the Faith of a Scholar.* Philadelphia: Westminster Press, 1949. Pp. 145–159.

From *The Self and the Dramas of History,* by Reinhold Niebuhr. Copyright 1955 by Charles Scribner's Sons. Reprinted by permission of the publisher. Reprinted by permission of Faber and Faber from *The Self and the Dramas of History* by R. Niebuhr.

PART II

Bertrand Russell, "A Free Man's Workshop." From B. Russell, *Why I Am Not A Christian.* New York: Simon and Schuster, 1957. Copyright © 1957 by Allen and Unwin. Reprinted by permission of Simon and Schuster, a Division of Gulf and Western Corporation. Reprinted by permission of George Allen and Unwin.

Julian Huxley, "I Believe." From C. Fadiman (ed.), *I Believe.* New York: Simon and Schuster, 1939, pp. 127–136. Reprinted by permission of A. D. Peters & Co., Ltd.

From *The Myth of Sisyphus and Other Essays,* by Albert Camus, translated by Justin O'Brien. Copyright © 1955 by Alfred A. Knopf, Inc. Reprinted by permission of Alfred A. Knopf.

Kurt Baier, "The Meaning of Life." Inaugural Lecture delivered at the Canberra University College, 1957. Copyright 1957 by Kurt Baier. Reprinted by permission of the author.

"Meaning and Value of Life" by Paul Edwards. Reprinted with permission of
the publisher from *The Encyclopedia of Philosophy*, Paul Edwards,
Editor in Chief. Volume 4, pages 467–477. Copyright © 1967 by
Macmillan, Inc.

Richard Taylor, "The Meaning of Life." Reprinted with permission of Macmillan Publishing Co., Inc. from *Good and Evil* by Richard Taylor.
Copyright © 1970 by Richard Taylor.

Thomas Nagel, "The Absurd." *The Journal of Philosophy*, Vol. LXIII, No. 20,
1971. Pp. 716–727. Reprinted by permission of the editor and the
author.

E. D. Klemke, "Living Without Appeal: An Affirmative Philosophy of Life."
E. D. Klemke, *Reflections and Perspectives*. The Hague: Mouton Publishers, 1974. Chapter 8, pp. 96–109. Reprinted by permission of the
publisher.

                                    PART III

Kai Nielsen, "Linguistic Philosophy and 'The Meaning of Life'." *Cross-
Currents* (Summer, 1964), pp. 313–334. This revised version is reprinted by permission of the editor and the author.

John Wisdom, "The Meanings of the Question of Life." In J. Wisdom,
Paradox and Discovery. Oxford: Basil Blackwell, 1965. Pp. 38–42.
Reprinted by permission of the publisher and the author.

R. W. Hepburn, "Questions About the Meaning of Life." *Religious Studies*,
Vol. 1 (1965), pp. 125–140. Copyright 1965 by Cambridge University Press. Reprinted by permission of Cambridge University Press.

"Why?" by Paul Edwards. Reprinted with permission of the publisher from
The Encyclopedia of Philosophy, Paul Edwards, Editor in Chief. Volume, 8, pp. 296–302. Copyright © 1967 by Macmillan, Inc.

R. M. Hare, "Nothing Matters." Reprinted from R. M. Hare, *Applications of
Moral Philosophy* (pp. 32–39), London: Macmillan, 1972, with the
author's permission and by permission of Macmillan, London and
Basingstoke. Copyright © R. M. Hare 1972.

W. D. Joske, "Philosophy and the Meaning of Life." *Australasian Journal of
Philosophy*, Vol. 52, no. 2, 1974, pp. 93–104. Reprinted by permission
of the editor.

Bertrand Russell. Prologue to *The Autobiography of Bertrand Russell*. London:
George Allen and Unwin, 1967. Reprinted with the permission of
George Allen and Unwin.

Contents

Part III *Questioning the Question*

THE MEANING OF LIFE

INTRODUCTION

The Question of the Meaning of Life

E. D. Klemke

In his work, "A Confession," Tolstoy gives an account of how, when he was fifty and at the height of his career, he became deeply distressed by the conviction that life was meaningless. He wrote:

> Five years ago, something very strange began to happen with me; I was overcome by minutes at first of perplexity and then an arrest of life, as though I did not know how to live or what to do, and I lost myself and was dejected. But that passed and I continued to live as before. Then those moments of perplexity were repeated oftener and oftener, and always in one and the same form. These arrests of life found their expression in ever the same question: "Why? Well, and then?"
>
> At first I thought that those were simply aimless, inappropriate questions. It seemed to me that that was all well known and that if I wanted to busy myself with their solution, it would not cost me much labour,— that now I had no time to attend to them, but that if I wanted to I should find the proper answers. But the questions began to repeat themselves oftener and oftener, answers were demanded more and more persistently, and like dots that fall on the same spot, these questions, without any answers, thickened into one black blotch. . . .
>
> I felt that what I was standing on had given way, that I had no foundation to stand on, that that which I lived by no longer existed, and that I had nothing to live by. . . .
>
> "Well, I know," I said to myself, "all which science wants so persistently to know, but there is no answer to the question about the meaning of life". . . .[1]

1. From L. Tolstoy, *My Confession,* trans. Leo Weiner (London: J. M. Dent & Sons, 1905), passim.

3

Perhaps almost every sensitive and reflective person has had at least some moments when similar fears and questions have arisen in his or her life. Perhaps the experiences were not as extreme as Tolstoy's was, but they have nevertheless been troublesome. And surely almost everyone has at some time asked: What is the meaning of life? Is there any meaning at all? What is it all about? What is the point of it all? It seems evident, then, that the question of the meaning of life is one of the most important questions of all.

At least one writer has maintained that it is *the* most urgent question of them all. In *The Myth of Sisyphus,* Camus writes:

> There is but one truly serious philosophical problem, and that is suicide. Judging whether life is or is not worth living amounts to answering the fundamental question of philosophy. All the rest—whether the world has three dimensions, whether the mind has nine or twelve categories—come afterward. These are games; one must first answer. . . .
>
> If I ask myself how to judge that this question is more urgent than that, I reply that one judges by the actions it entails. I have never seen anyone die for the ontological argument [for the existence of a god]. Galileo, who held a scientific truth of great importance, abjured it with the greatest of ease as soon as it endangered his life. In a certain sense he did right. That truth was not worth the stake. Whether the earth or the sun revolved around the other is a matter of profound indifference. To tell the truth, it is a futile question. On the other hand, I see many people die because they judge that life is not worth living. I see others paradoxically getting killed for the ideas or illusions that give them a reason for living (what is called a reason for living is also an excellent reason for dying). I therefore conclude that the meaning of life is the most urgent of questions.[2]

However we may rank the question—as the most urgent of all or as one of the most urgent of all—most of us do find the question to be one that merits our serious attention. Part of its urgency stems from the fact that it is related to many other questions that face us in our daily lives. Many of the decisions we make with regard to careers, leisure time, moral dilemmas, and other matters, depend on how we answer the question of the meaning of life.

However, the question may mean several things. Let us attempt to distinguish some of them. The question "What is the meaning of life?" may mean: (1) Why does the universe exist? Why is there something rather than nothing? Is there some plan for the whole universe? (2) Why

2. A. Camus, *The Myth of Sisyphus,* trans. J. O'Brien (N.Y.: Alfred A. Knopf, Inc., 1955), p. 1.

do humans (in general) exist? Do they exist for some purpose? If so, what is it? (3) Why do *I* exist? Do I exist for some purpose? If so, how am I to find what it is? If not, how can life have any significance or value?

I do not mean to suggest that these are rigidly distinct questions. They are obviously interrelated. Hence we may interpret the question "What is the meaning of life?" broadly so that it can include any one or any two or all three of these questions. In so doing, we will be following common usage. Perhaps most of those who are deeply concerned with the question interpret it mainly in sense (3). However, there are others who include either (1) or (2) or both along with (3). Again, we need not be concerned with specifying one of the interpretations as the "correct" one. Rather let us recognize that all of them are often involved when one asks about the meaning of life.

In turning to possible answers to the question of the meaning of life, we find different approaches or stances. These are:

    I. The theistic answer
   II. The nontheistic (or antitheistic) alternative
  III. The approach that questions the meaningfulness of the question

These three different approaches are represented by the readings found in Parts I, II and III, respectively, of this volume. Perhaps a brief summary of each would be desirable at this point.

I. According to the theistic answer, the meaning of life is found in the existence of a god—a supremely benevolent and all-powerful being, a being transcendent to the natural universe but who created the universe and fashioned man in his image and endowed him with a preordained purpose. In this view, without the existence of God, or at least without faith in God, life has no meaning or purpose and hence is not worth living. It is difficult to find many works in which this position is explicitly defended. However, it is a view that is held widely by religious believers, and it is apparent that in many cases it is a basis (or part of the basis) for religious belief. Along with Tolstoy's *Confession* the other essays found in Part I of this volume come close to providing a direct defense of this view.

II. The nontheistic (or humanistic) alternative, of course, denies the claim that the meaning of life is dependent upon the existence of a god. According to this alternative, since there is no good reason to believe in the existence of a transcendent god, there is no good reason to believe that life has any objective meaning or purpose—that is, any meaning that is dependent upon anything outside of the natural universe. Rather, the meaning of life, if there is one, must be found within the natural universe. Some adherents of this view go on to claim that there is no good reason to think life has any meaning in *any* objective sense, but there is good reason to believe that it can nevertheless have meaning in a subjective sense. In other words, it is up to each individual to fashion or create his or her own meaning by virtue of his own consciousness and creative activity. A defense of this position is found in the editor's own essay and in other essays in Part II of this volume.

III. There is yet a third approach to the question of the meaning of life. According to this approach, the question "What is the meaning of life?" is a peculiar or at least an ambiguous one. What kind of question are we asking when we ask about the meaning of life? There are some who think that, upon analysis of such terms as "meaning," "purpose," "value," etc., the question of the meaning of life turns out to be cognitively meaningless. However there are others who reject that claim and take the opposing view. According to them, the various questions that make up the larger question of the meaning of life can be given an interpretation that renders them to be intelligible and cognitively significant. Various aspects of and perspectives on this position are presented in the essays in Part III of this volume.

It is my view that the question of the meaning of life is a significant and important one and that it can be given an answer. I believe that the answer does not rest on any theistic or metaphysical assumptions. Since my defense of this position has been given in my essay, "Living Without Appeal: An Affirmative Philosophy of Life," contained in this volume, I will not repeat what I have said there except to say this: It will be recalled that, in the passage quoted above, Camus maintains that "Judging whether life is or is not worth living amounts to answering the fundamental question of philosophy." The answer I have given to that question is: There is no formula or slogan by which to guarantee that life will be worth living; but I believe that it *can* be worth living. I believe that I shall die dead, but that in the meantime, I can find a genuine meaning and purpose in life that makes it worth living.

There are some who will object: But how can life have any meaning or worth or value if it must come to an end? Sir Karl Popper has given the best answer I know of to that question. He writes: "There are those who think that life is valueless because it comes to an end. They fail to see that the opposite argument might also be proposed: that if there were no end to life, life would have no value; that it is, in part, the ever-present danger of losing it which helps to bring home to us the value of life."[3]

---

3. Sir Karl Popper, "How I See Philosophy," in A. Mercier and M. Svilar (eds.), *Philosophers on Their Own Work*, Vol. 3 (Berne and Frankfurt am Main: Peter Lang, 1977), p. 148.

PART I

THE THEISTIC ANSWER

❧ I ❦

LEO TOLSTOY

My Confession

Although I regarded authorship as a waste of time, I continued to write during those fifteen years. I had tasted of the seduction of authorship, of the seduction of enormous monetary remunerations and applauses for my insignificant labour, and so I submitted to it, as being a means for improving my material condition and for stifling in my soul all questions about the meaning of my life and life in general.

In my writings I advocated, what to me was the only truth, that it was necessary to live in such a way as to derive the greatest comfort for oneself and one's family.

Thus I proceeded to live, but five years ago something very strange began to happen with me: I was overcome by minutes at first of perplexity and then of an arrest of life, as though I did not know how to live or what to do, and I lost myself and was dejected. But that passed, and I continued to live as before. Then those minutes of perplexity were repeated oftener and oftener, and always in one and the same form. These arrests of life found their expression in ever the same questions: "Why? Well, and then?"

At first I thought that those were simply aimless, inappropriate questions. It seemed to me that that was all well known and that if I ever wanted to busy myself with their solution, it would not cost me much labour,—that now I had no time to attend to them, but that if I wanted to I should find the proper answers. But the questions began to repeat themselves oftener and oftener, answers were demanded more and more persistently, and, like dots that fall on the same spot, these questions, without any answers, thickened into one black blotch.

There happened what happens with any person who falls ill with a mortal internal disease. At first there appear insignificant symptoms of indisposition, to which the patient pays no attention; then these symptoms are repeated more and more frequently and blend into one temporally in-

9

divisible suffering. The suffering keeps growing, and before the patient has had time to look around, he becomes conscious that what he took for an indisposition is the most significant thing in the world to him,—is death.

The same happened with me. I understood that it was not a passing indisposition, but something very important, and that, if the questions were going to repeat themselves, it would be necessary to find an answer for them. And I tried to answer them. The questions seemed to be so foolish, simple, and childish. But the moment I touched them and tried to solve them, I became convinced, in the first place, that they were not childish and foolish, but very important and profound questions in life, and, in the second, that, no matter how much I might try, I should not be able to answer them. Before attending to my Samára estate, to my son's education, or to the writing of a book, I ought to know why I should do that. So long as I did not know why, I could not do anything. I could not live. Amidst my thoughts of farming, which interested me very much during that time, there would suddenly pass through my head a question like this: "All right, you are going to have six thousand desyatínas of land in the Government of Samára, and three hundred horses,—and then?" And I completely lost my senses and did not know what to think farther. Or, when I thought of the education of my children, I said to myself: "Why?" Or, reflecting on the manner in which the masses might obtain their welfare, I suddenly said to myself: "What is that to me?" Or, thinking of the fame which my works would get me, I said to myself: "All right, you will be more famous than Gógol, Púshkin, Shakespeare, Molière, and all the writers in the world,—what of it?" And I was absolutely unable to make any reply. The questions were not waiting, and I had to answer them at once; if I did not answer them, I could not live.

I felt that what I was standing on had given way, that I had no foundation to stand on, that that which I lived by no longer existed, and that I had nothing to live by. . . .

All that happened with me when I was on every side surrounded by what is considered to be complete happiness. I had a good, loving, and beloved wife, good children, and a large estate, which grew and increased without any labour on my part. I was respected by my neighbours and friends, more than ever before, was praised by strangers, and, without any self-deception, could consider my name famous. With all that, I was not deranged or mentally unsound,—on the contrary, I was in full command of my mental and physical powers, such as I had rarely met with in people of my age: physically I could work in a field, mowing, without falling behind a peasant; mentally I could work from eight to ten hours in succession, without experiencing any consequences from the strain. And while in such condition I arrived at the conclusion that I could not live, and, fearing death, I had to use cunning against myself, in order that I might not take my life.

This mental condition expressed itself to me in this form: my life is a stupid, mean trick played on me by somebody. Although I did not recognize that "somebody" as having created me, the form of the conception that some one had played a mean, stupid trick on me by bringing me into the world was the most natural one that presented itself to me.

Involuntarily I imagined that there, somewhere, there was somebody who was now having fun as he looked down upon me and saw me, who had lived for thirty or forty years, learning, developing, growing in body and mind, now that I had become strengthened in mind and had reached that summit of life from which it lay all before me, standing as a complete fool on that summit and seeing clearly that there was nothing in life and never would be. And that was fun to him—

But whether there was or was not that somebody who made fun of me, did not make it easier for me. I could not ascribe any sensible meaning to a single act, or to my whole life. I was only surprised that I had not understood that from the start. All that had long ago been known to everybody. Sooner or later there would come diseases and death (they had come already) to my dear ones and to me, and there would be nothing left but stench and worms. All my affairs, no matter what they might be, would sooner or later be forgotten, and I myself should not exist. So why should I worry about all these things? How could a man fail to see that and live,— that was surprising! A person could live only so long as he was drunk; but the moment he sobered up, he could not help seeing that all that was only a deception, and a stupid deception at that! Really, there was nothing funny and ingenious about it, but only something cruel and stupid.

Long ago has been told the Eastern story about the traveller who in the steppe is overtaken by an infuriated beast. Trying to save himself from the animal, the traveller jumps into a waterless well, but at its bottom he sees a dragon who opens his jaws in order to swallow him. And the unfortunate man does not dare climb out, lest he perish from the infuriated beast, and does not dare jump down to the bottom of the well, lest he be devoured by the dragon, and so clutches the twig of a wild bush growing in a cleft of the well and holds on to it. His hands grow weak and he feels that soon he shall have to surrender to the peril which awaits him at either side; but he still holds on and sees two mice, one white, the other black, in even measure making a circle around the main trunk of the bush to which he is clinging, and nibbling at it on all sides. Now, at any moment, the bush will break and tear off, and he will fall into the dragon's jaws. The traveller sees that and knows that he will inevitably perish; but while he is still clinging, he sees some drops of honey hanging on the leaves of the bush, and so reaches out for them with his tongue and licks the leaves. Just so I hold on to the branch of life, knowing that the dragon of death is waiting inevitably for me, ready to tear me to pieces, and I cannot understand why I have fallen on such suffering. And I try to lick that honey which used to give me pleasure; but now it no longer gives me joy, and the white and the black mouse day and night nibble at the branch to which I am holding on. I clearly see the dragon, and the honey is no longer sweet to me. I see only the inevitable dragon and the mice, and am unable to turn my glance away from them. That is not a fable, but a veritable, indisputable, comprehensible truth.

The former deception of the pleasures of life, which stifled the terror of the dragon, no longer deceives me. No matter how much one should say to me, "You cannot understand the meaning of life, do not think, live!" I am unable to do so, because I have been doing it too long before. Now I cannot help seeing day and night, which run and lead me up to death. I see that alone, because that alone is the truth. Everything else is a lie.

The two drops of honey that have longest turned my eyes away from the cruel truth, the love of family and of authorship, which I have called an art, are no longer sweet to me.

"My family—" I said to myself, "but my family, my wife and children, they are also human beings. They are in precisely the same condition that I am in: they must either live in the lie or see the terrible truth. Why should they live? Why should I love them, why guard, raise, and watch them? Is it for the same despair which is in me, or for dulness of perception? Since I love them, I cannot conceal the truth from them,— every step in cognition leads them up to this truth. And the truth is death."

"Art, poetry?" For a long time, under the influence of the success of human praise, I tried to persuade myself that that was a thing which could be done, even though death should come and destroy everything, my deeds, as well as my memory of them; but soon I came to see that that, too, was a deception. It was clear to me that art was an adornment of life, a decoy of life. But life lost all its attractiveness for me. How, then, could I entrap others? So long as I did not live my own life, and a strange life bore me on its waves; so long as I believed that life had some sense, although I was not able to express it,—the reflections of life of every description in poetry and in the arts afforded me pleasure, and I was delighted to look at life through this little mirror of art; but when I began to look for the meaning of life, when I experienced the necessity of living myself, that little mirror became either useless, superfluous, and ridiculous, or painful to me. I could no longer console myself with what

I saw in the mirror, namely, that my situation was stupid and desperate. It was all right for me to rejoice so long as I believed in the depth of my soul that life had some sense. At that time the play of lights—of the comical, the tragical, the touching, the beautiful, the terrible in life—afforded me amusement. But when I knew that life was meaningless and terrible, the play in the little mirror could no longer amuse me. No sweetness of honey could be sweet to me, when I saw the dragon and the mice that were nibbling down my support. . . .

In my search after the question of life I experienced the same feeling which a man who has lost his way in the forest may experience.

He comes to a clearing, climbs a tree, and clearly sees an unlimited space before him; at the same time he sees that there are no houses there, and that there can be none; he goes back to the forest, into the darkness, and he sees darkness, and again there are no houses.

Thus I blundered in this forest of human knowledge, between the clearings of the mathematical and experimental sciences, which disclosed to me clear horizons, but such in the direction of which there could be no house, and between the darkness of the speculative sciences, where I sunk into a deeper darkness, the farther I proceeded, and I convinced myself at last that there was no way out and could not be.

By abandoning myself to the bright side of knowledge I saw that I only turned my eyes away from the question. No matter how enticing and clear the horizons were that were disclosed to me, no matter how enticing it was to bury myself in the infinitude of this knowledge, I comprehended that these sciences were the more clear, the less I needed them, the less they answered my question.

"Well, I know," I said to myself, "all which science wants so persistently to know, but there is no answer to the question about the meaning of my life." But in the speculative sphere I saw that, in spite of the fact that the aim of the knowledge was directed straight to the answer of my question, or because of that fact, there could be no other answer than what I was giving to myself: "What is the meaning of my life?"—"None." Or, "What will come of my life?"—"Nothing." Or, "Why does everything which exists exist, and why do I exist?"—"Because it exists."

Putting the question to the one side of human knowledge, I received an endless quantity of exact answers about what I did not ask: about the chemical composition of the stars, about the movement of the sun toward the constellation of Hercules, about the origin of species and of man, about the forms of infinitely small, imponderable particles of ether; but the answer in this sphere of knowledge to my question what the meaning of my life was, was always: "You are what you call your life; you are a temporal, accidental conglomeration of particles. The interrelation, the change of these particles, produces in you that which you call life. This congeries will last for some time; then the interaction of these particles will cease, and that which you call life and all your questions will come to an end. You are an accidentally cohering globule of something. The globule is fermenting. This fermentation the globule calls its life. The globule falls to pieces, and all fermentation and all questions will come to an end." Thus the clear side of knowledge answers, and it cannot say anything else, if only it strictly follows its principles.

With such an answer it appears that the answer is not a reply to the question. I want to know the meaning of my life, but the fact that it is a particle of the infinite not only gives it no meaning, but even destroys every possible meaning.

Those obscure transactions, which this side of the experimental, exact science has with speculation, when it says that the meaning of life consists in evolution and the coöperation with this evolution, because of their obscurity and inexactness cannot be regarded as answers.

The other side of knowledge, the speculative, so long as it sticks strictly to its fundamental principles in giving a direct answer to the question, everywhere and at all times has answered one and the same: "The world is something infinite and incomprehensible. Human life is an incomprehensible part of this incomprehensible all. . . ."

I lived for a long time in this madness, which, not in words, but in deeds, is particularly characteristic of us, the most liberal and learned of men. But, thanks either to my strange, physical love for the real working class, which made me understand it and see that it is not so stupid as we suppose, or to the sincerity of my conviction, which was that I could know nothing and that the best that I could do was to hang myself,—I felt that if I wanted to live and understand the meaning of life, I ought naturally to look for it, not among those who had lost the meaning of life and wanted to kill themselves, but among those billions departed and living men who had been carrying their own lives and ours upon their shoulders. And I looked around at the enormous masses of deceased and living men,—not learned and wealthy, but simple men,—and I saw something quite different. I saw that all these billions of men that lived or had lived, all, with rare exceptions, did not fit into my subdivisions,* and that I

---

* In a passage omitted here, Tolstoy characterized four attitudes that people have towards life: living in ignorance of the problem of the meaning of life; ignoring it and trying to attain as much pleasure as possible; admitting that life is meaningless and committing suicide; admitting that life is meaningless but continuing to live aimlessly. (Ed.)

could not recognize them as not understanding the question, because they themselves put it and answered it with surprising clearness. Nor could I recognize them as Epicureans, because their lives were composed rather of privations and suffering than of enjoyment. Still less could I recognize them as senselessly living out their meaningless lives, because every act of theirs and death itself was explained by them. They regarded it as the greatest evil to kill themselves. It appeared, then, that all humanity was in possession of a knowledge of the meaning of life, which I did not recognize and which I contemned. It turned out that rational knowledge did not give any meaning to life, excluded life, while the meaning which by billions of people, by all humanity, was ascribed to life was based on some despised, false knowledge.

The rational knowledge in the person of the learned and the wise denied the meaning of life, but the enormous masses of men, all humanity, recognized this meaning in an irrational knowledge. This irrational knowledge was faith, the same that I could not help but reject. That was God as one and three, the creation in six days, devils and angels, and all that which I could not accept so long as I had not lost my senses.

My situation was a terrible one. I knew that I should not find anything on the path of rational knowledge but the negation of life, and there, in faith, nothing but the negation of reason, which was still more impossible than the negation of life. From the rational knowledge it followed that life was an evil and men knew it,—it depended on men whether they should cease living, and yet they lived and continued to live, and I myself lived, though I had known long ago that life was meaningless and an evil. From faith it followed that, in order to understand life, I must renounce reason, for which alone a meaning was needed.

There resulted a contradiction, from which there were two ways out: either what I called rational was not so rational as I had thought; or that which to me appeared irrational was not so irrational as I had thought. And I began to verify the train of thoughts of my rational knowledge.

In verifying the train of thoughts of my rational knowledge, I found that it was quite correct. The deduction that life was nothing was inevitable; but I saw a mistake. The mistake was that I had not reasoned in conformity with the question put by me. The question was, "Why should I live?" that is, "What real, indestructible essence will come from my phantasmal, destructible life? What meaning has my finite existence in this infinite world?" And in order to answer this question, I studied life.

The solutions of all possible questions of life apparently could not satisfy me, because my question, no matter how simple it appeared in the beginning, included the necessity of explaining the finite through the infinite, and vice versa.

I asked, "What is the extra-temporal, extra-causal, extra-spatial meaning of life?" But I gave an answer to the question, "What is the temporal, causal, spatial meaning of my life?" The result was that after a long labour of mind I answered, "None."

In my reflections I constantly equated, nor could I do otherwise, the finite with the finite, the infinite with the infinite, and so from that resulted precisely what had to result: force was force, matter was matter, will was will, infinity was infinity, nothing was nothing,—and nothing else could come from it.

There happened something like what at times takes place in mathematics: you think you are solving an equation, when you have only an identity. The reasoning is correct, but you receive as a result the answer: $a = a$, or $x = x$, or $o = o$. The same happened with my reflection in respect to the question about the meaning of my life. The answers given by all science to that question are only identities.

Indeed, the strictly scientific knowledge, that knowledge which, as Descartes did, begins with a full doubt in everything, rejects all knowledge which has been taken on trust, and builds everything anew on the laws of reason and experience, cannot give any other answer to the question of life than what I received,—an indefinite answer. It only seemed to me at first that science gave me a positive answer,—Schopenhauer's answer: "Life has no meaning, it is an evil." But when I analyzed the matter, I saw that the answer was not a positive one, but that it was only my feeling which expressed it as such. The answer, strictly expressed, as it is expressed by the Brahmins, by Solomon, and by Schopenhauer, is only an indefinite answer, or an identity, $o = o$, life is nothing. Thus the philosophical knowledge does not negate anything, but only answers that the question cannot be solved by it, that for philosophy the solution remains insoluble.

When I saw that, I understood that it was not right for me to look for an answer to my question in rational knowledge, and that the answer given by rational knowledge was only an indication that the answer might be got if the question were differently put, but only when into the discussion of the question should be introduced the question of the relation of the finite to the infinite. I also understood that, no matter how irrational and monstrous the answers might be that faith gave, they had this advantage that they introduced into each answer the relation of the finite to the infinite, without which there could be no answer.

No matter how I may put the question, "How must I live?" the answer is, "According to God's law." "What real result will there be from my life?"—"Eternal torment or eternal bliss." "What is the meaning which is not destroyed by death?"—"The union with infinite God, paradise."

Thus, outside the rational knowledge, which had to me appeared as the only one, I was inevitably led to recognize that all living humanity had a certain other irrational knowledge, faith, which made it possible to live.

All the irrationality of faith remained the same for me, but I could not help recognizing that it alone gave to humanity answers to the questions of life, and, in consequence of them, the possibility of living.

The rational knowledge brought me to the recognition that life was meaningless,—my life stopped, and I wanted to destroy myself. When I looked around at people, at all humanity, I saw that people lived and asserted that they knew the meaning of life. I looked back at myself: I lived so long as I knew the meaning of life. As to other people, so even to me, did faith give the meaning of life and the possibility of living.

Looking again at the people of other countries, contemporaries of mine and those passed away, I saw again the same. Where life had been, there faith, ever since humanity had existed, had given the possibility of living, and the chief features of faith were everywhere one and the same.

No matter what answers faith may give, its every answer gives to the finite existence of man the sense of the infinite,—a sense which is not destroyed by suffering, privation, and death. Consequently in faith alone could we find the meaning and possibility of life. What, then, was faith? I understood that faith was not merely an evidence of things not seen, and so forth, not revelation (that is only the description of one of the symptoms of faith), not the relation of man to man (faith has to be defined, and then God, and not first God, and faith through him), not merely an agreement with what a man was told, as faith was generally understood,— that faith was the knowledge of the meaning of human life, in consequence of which man did not destroy himself, but lived. Faith is the power of life. If a man lives he believes in something. If he did not believe that he ought to live for some purpose, he would not live. If he does not see and understand the phantasm of the finite, he believes in that finite; if he understands the phantasm of the finite, he must believe in the infinite. Without faith one cannot live. . . .

In order that all humanity may be able to live, in order that they may continue living, giving a meaning to life, they, those billions, must have another, a real knowledge of faith, for not the fact that I, with Solomon and Schopenhauer, did not kill myself convinced me of the existence of faith, but that these billions had lived and had borne us, me and Solomon, on the waves of life.

Then I began to cultivate the acquaintance of the believers from among the poor, the simple and unlettered folk, of pilgrims, monks, dissenters, peasants. The doctrine of these people from among the masses was also the Christian doctrine that the quasi-believers of our circle professed. With the Christian truths were also mixed in very many superstitions, but there was this difference: the superstitions of our circle were quite unnecessary to them, had no connection with their lives, were only a kind of an Epicurean amusement, while the superstitions of the believers from among the labouring classes were to such an extent blended with their life that it would have been impossible to imagine it without these superstitions,—it was a necessary condition of that life. I began to examine closely the lives and beliefs of these people, and the more I examined them, the more did I become convinced that they had the real faith, that their faith was necessary for them, and that it alone gave them a meaning and possibility of life. In contradistinction to what I saw in our circle, where life without faith was possible, and where hardly one in a thousand professed to be a believer, among them there was hardly one in a thousand who was not a believer. In contradistinction to what I saw in our circle, where all life passed in idleness, amusements, and tedium of life, I saw that the whole life of these people was passed in hard work, and that they were satisfied with life. In contradistinction to the people of our circle, who struggled and murmured against fate because of their privations and their suffering, these people accepted diseases and sorrows without any perplexity or opposition, but with the calm and firm conviction that it was all for good. In contradistinction to the fact that the more intelligent we are, the less do we understand the meaning of life and the more do we see a kind of a bad joke in our suffering and death, these people live, suffer, and approach death, and suffer in peace and more often in joy. In contradistinction to the fact that a calm death, a death without terror or despair, is the greatest exception in our circle, a restless, insubmissive, joyless death is one of the greatest exceptions among the masses. And of such people, who are deprived of everything which for Solomon and for me constitutes the only good of life, and who withal experience the greatest happiness, there is an enormous number. I cast a broader glance about me. I examined the life of past and present vast masses of men, and I saw people who in like manner had understood the meaning of life, who had known how to live and die, not two, not three, not ten, but hundreds, thousands, millions. All of them, infinitely diversified as to habits, intellect, culture, situation, all equally and quite contrary to my ignorance knew the meaning of life and of death, worked calmly, bore privations and suffering, lived and died, seeing in that not vanity, but good.

I began to love those people. The more I penetrated into their life, the life of the men now living, and the life of men departed, of whom I had read and heard, the more did I love them, and the easier it became for me to live. Thus I lived for about two years, and within me took place a trans-formation, which had long been working within me, and the germ of which had always been in me. What happened with me was that the life of our circle,—of the rich and the learned,—not only disgusted me, but even lost all its meaning. All our acts, reflections, sciences, arts,—all that appeared to me in a new light. I saw that all that was mere pampering of the appetites, and that no meaning could be found in it; but the life of all the working masses, of all humanity, which created life, presented itself to me in its real significance. I saw that that was life itself and that the meaning given to this life was truth, and I accepted it.

DAVID F. SWENSON

The Dignity
of Human Life

Man lives forward, but he thinks backward. As an active being, his task is to press forward to the things that are before, toward the goal where is the prize of the high calling. But as a thinking, active being, his forward movement is conditioned by a retrospect. If there were no past for a man, there could be no future; and if there were no future and no past, but only such an immersion in the present as is characteristic of the brute which perisheth, then there would be nothing eternal in human life, and everything distinctively and essentially human would disappear from our existence.

As a preparation for an existence in the present, the youth of a nation are trained in various skills and along devious lines, according to their capacities and circumstances, for the parts they are to play in existence; their natural talents are developed, some by extended periods of intellectual training, others for participation in various forms of business or technical training; but whatever be the ultimate end of the training, its purpose is to develop those latent powers they possess which will eventually prove of benefit to themselves or to others. But, in addition to this, which we may call a preparation for the external life, a something else is urgently needed, a something so fundamentally important that in its absence every other form of preparation is revealed as imperfect and incomplete, even ineffective and futile.

This so particularly indispensable something is a view of life, and a view of life is not acquired as a direct and immediate result of a course of study, the reading of books, or a communication of results. It is wholly a product of the individual's own knowledge of himself as an individual, of his individual capabilities and aspirations. A view of life is a principle of living, a spirit and an attitude capable of maintaining its unity and identity with itself in all of life's complexities and varying vicissitudes; and yet also capable of being declined, to use the terminology of the grammatical sciences, in all the infinite variety of cases that the language of life affords. Without this preparation the individual life is like a ship without a rudder, a bit of wreckage floating with the current to an uncomprehended destiny. A view of life is not objective knowledge, but subjective conviction. It is a part of a man's own self, the source whence the stream of his life issues. It is the dominant attitude of the spirit which gives to life its direction and its goal. This is why it cannot be directly communicated or conveyed, like an article of commerce, from one person to another. If a view of life were a body of knowledge about life, or a direct and immediate implication from such knowledge, it would be subject to objective communication and systematic instruction. But it is rather a personal expression of what a man essentially is in his own inmost self, and this cannot be learned by rote, or accepted at the hands of some external authority. Knowledge is the answer or answers that things give to the questions we ask of them; a view of life is the reply a person gives to the question that life asks of him. We begin life by scrutinizing our environment, ourselves playing the role of questioners and examiners and critics; but at a later moment, when the soul comes of age and is about to enter upon its majority, it learns that the tables have been turned and the roles reversed; from that moment it confronts a question, a searching and imperative question, in relation to which no evasion can avail, and to which no shifting of responsibility is possible.

In discussing the problem of a view of life which can give it meaning and dignity and worth, I am well aware that no one can acquire a view of life by listening to a speech. Nevertheless, a speech may serve the more modest purpose of stimulating a search, perhaps a more earnest search; and may render more articulate possibly the convictions of those who have already won some such conception, having made it their own by a heartfelt and spontaneous choice.

All men are endowed by nature with a desire for happiness—a principle so obvious as scarcely to need any explanation, and certainly no defense. A human life without happiness or hope of happiness is not a life, but rather a death in life. Happiness is life's vital fluid and the very breath of its nostrils. Happiness and life are so much one and the same thing that the more profoundly any life discovers happiness, the more significant and abundant is that life itself. This is also the principle of the Christian religion, which even dares to formulate the task of life as the duty of being happy; for does not the Apostle Paul say, "Rejoice . . . always: again I will say, Rejoice"? So deeply grounded in human nature is

the need for happiness, that the desire for it is not only universal and instinctive, but ineradicable and imperative. Man is made for happiness; an
essentially unhappy man has missed his goal, and has failed to realize his
humanity.

But for a thinking human being—and God made every man a thinker,
whatever may be our estimate of that which men make of themselves—for
a thinking human being, happiness cannot consist in the satisfaction of
momentary impulse, of blind feeling, of brute immediacy. A pleasant ab-
sorption in the present, oblivious of prospect or retrospect, careless of the
wider relations or the deeper truth of life, can be called happiness only on
the basis of frivolity and thoughtlessness. Just as life is not life unless it is
happy, so happiness is not happiness unless it can be justified. In order
really to be happiness it requires to be interpenetrated with a sense of
meaning, reason, and worth.

For the quest for happiness, like every other human quest, faces a
danger. The danger that confronts it is the possibility of error: the error
of permitting oneself to be lured into promising paths that lead to no goal,
and the error of coming to rest in hollow satisfactions and empty joys. It
is possible to believe oneself happy, to seem happy to oneself and to others,
and yet in reality to be plunged in the deepest misery; just as, on the
other hand, it is possible to stand possessed of the greatest treasure, and
yet, in thoughtlessness, to imagine oneself destitute, and through that very
thoughtlessness not only neglect and ignore but actually deprive oneself of
what one already has. The basic problem of life, the question in response
to which a view of life seeks to propound an answer, may therefore be formulated as follows: What is that happiness which is also a genuine and
lasting good? In what does it consist, and how may it be attained?

There exists an ancient handbook, an *Art of Rhetoric,* compiled for
the guidance and information of orators and other public speakers, written
by one of the greatest of Greek philosophers. In this handbook the author formulates the commonly prevailing conceptions of happiness as
among the things useful for public speakers to know. This textbook is almost twenty-five hundred years old, and the views it presents on this subject may therefore be expected to seem childish in the light of our greater
insight and extraordinary progress in all things. Nevertheless, let us note
them in passing, if only for the sake of seeing how far we have advanced
beyond them. Happiness is said to be commonly defined as independence
of life, as prosperity with virtue, as comfortable circumstances with security, or as the enjoyment of many possessions, together with the power to
keep and defend them. Its constituent elements are noble birth, wealth,
many good and influential friends, bodily health and beauty, talents and
capacities, good fortune, honors, and lastly virtue. We readily perceive

how strange and old-fashioned these conceptions are, how foreign to all our modern and enlightened notions. I shall therefore subjoin a more up-to-date consideration of the same subject, derived from a very modern author writing in a journal of today. The author raises the question as to what circumstances and conditions have the power to make him feel really alive, tingling with vitality, instinct with the joy of living. He submits a long list including a variety of things, of which I shall quote the chief: the sense of health; successful creative work, like writing books; good food and drink; pleasant surroundings; praise, not spread on too thick; friends and their company; beautiful things, books, music; athletic exercises and sports; daydreaming; a good fight in a tolerably decent cause; the sense of bodily danger escaped; the consciousness of being a few steps ahead of the wolf of poverty. His social ideal is a community where beauty abounds, where the fear of want is absent, where a man may dress as he pleases, do the work that suits him best; a community where arts and letters flourish, where abundant leisure breaks the remorseless pace of ceaseless work, and, lastly, for those who value religion, a church is provided, with a great nave and a great organ and the sound of vespers across the evening fields. So speaks our modern writer. And now that I have juxtaposed these two accounts, I have to confess to the strange feeling that, despite the interval of more than two thousand years between them, they sound unexpectedly alike, even to the generous inclusion of a place for morality and religion as not an entirely negligible factor in promoting the good and happy life. How strange to find such a similarity! Can it be that after all that has been said and written about the revolutionary and radical changes introduced into life by modern science, modern invention, and modern industry, the influence of the steam engine and the printing press, the telegraph and the radio, the automobile and the airplane, together with the absolutely devastating discoveries of astronomers—can it be, in spite of all this, that the current conceptions of life and its meaning have remained essentially unchanged? Is not this a remarkable testimony to the profound inner resemblance to one another, despite all changes of form and circumstance, exhibited by the countless generations of men, both in their wisdom and in their folly?

However that may be, I do not think that anyone will deny that such views as these are widely held, and constitute the view of life perhaps of the majority of men. The testimony to their prevalence is not merely the articulate confession of the tongue, but the no less revealing though inarticulate direction of the life. I hope not to be misunderstood. The present speaker is human enough to find these objectives, or the majority of them, not only natural but inviting; he finds them desirable, and is by no means schooled in any heroic or stoic indifference to the goods that have rightly been called external, rooted as they are in fortunate circumstances and special privilege. But there are serious difficulties in the way of constructing a view of life out of such considerations.

The constituents of happiness are in both cases a multiplicity of things. As Aristotle so simply says, virtue alone will not make a man happy, but he needs also goods and friends. But the self which sets its heart upon any such multiplicity of external goods, which lives in them and by them and for them, dependent upon them for the only happiness it knows—such a self is captive to the diverse world of its desires. It belongs to the world and does not own itself. It is not in the deepest sense a self, since it is not free and is not a unity. The manifold conditions of its happiness split the self asunder; no ruling passion dominates its life; no concentration gives unity to the personality and single-mindedness to the will. Its name is legion, and its nature is double-mindedness. And if some one thing, like wealth or power, is made the single ambition of an exceptional life, it still remains true that such things are only apparently single; in reality they are various and manifold. The soul that lives in them is torn by diverse impulses, is drawn in many different directions at once, and cannot find the peace which comes only from single-minded devotion, from the pursuit of an end which is intrinsically and genuinely one.

Reflection discovers yet another difficulty in connection with such views of life. Whoever seeks his happiness in external conditions, of whatever sort, seeks it in that which is in its essential nature precarious. He presumes upon the realization of conditions which are not intrinsic to him, or within his control. This happiness is subject to the law of uncertainty, to the qualification of an unyielding, mysterious *perhaps*. Here lurks the possibility of despair. Give a man the full satisfaction of his wishes and ambitions, and he deems himself happy; withdraw from him the smile of fortune's favor, and disappoint his expectation and his hope, and he will be plunged into despair. The shift from happiness to unhappiness in such a life is every moment imminent. And therefore its despair is latent even in its happiness, and discord lurks imminent in the soul's most beautiful music; its presence is masked only by a brutish self-satisfaction, the habit of thoughtlessness, breathless haste in trifling errands, and darkness in the soul's deepest ground, each and all miserable defenses indeed against the enemy within the gates.

A third consideration. Wealth and power and the like, even bodily health and beauty of person, are not in the strictest sense intrinsic values, but rather representative and comparative, conditional and hypothetical. Money is good—if I have learned how to use it; and so with power and influence, health and strength. But in themselves these things are abstract and neutral, and no man can truthfully say whether the acquirement of them in any individual case will work more good than harm. This is a consideration which applies to nearly every item of what we call progress. A new discovery or invention, like the printing press, promises radically to improve life, to secure for us a hitherto undreamed-of happiness and well-being. We hail it with enthusiasm as inaugurating a new era, and in the distance we descry the dawn of a millennial day. A century or so passes. What then? Why, all the old difficulties and problems, all the old dissatisfactions and complaints, the very difficulties and problems that were to be solved by the new invention, are seen to be still with us, having but slightly changed their outward form and habitat. In addition, the new improvement that was to usher in the millennium, is seen to be the source of so many and so serious abuses (consider the abuse of the printing press!) that the best minds of the race have to be concentrated upon the problem of finding a remedy for these abuses, and keeping them under some sort of control. And so also in the individual life. Every access of power and prosperity, and of outward comfort, brings with it its own griefs and dangers. Every such improvement is only potentially a good, as it is also potentially an evil. It is a mere quantity whose qualitative meaning is indeterminate, awaiting the seal of something else, something from within the soul itself, in order to determine its final significance for weal or woe, for happiness or unhappiness.

Lastly, it must be pointed out that the conditions of happiness as conceived in all such views of life, inevitably imply a privileged status for the happy individual. They rest upon differential capabilities and exceptionally fortunate circumstances. To choose them as the end and aim of life constitutes an injury to the mass of men who are not so privileged. This one thought alone is of so arresting a quality as to give the deepest concern to every man who has the least trace of human sympathy and human feeling. I hope I have a soul not entirely a stranger to happy admiration; I know I feel moved to bend low in respect before exceptional talent and performance, and that I am eager to honor greatness and genius wherever I have the gift to understand it. And I am not so unfeeling as to refuse a tribute of sympathetic joy to those who rejoice in fortune's favors and bask in the smiles of outward success. But as the fundamental source of inspiration of my life, I need something that is not exclusive and differential, but inclusive and universal. I require to drink from a spring at which all men may refresh themselves; I need an aim that reconciles me to high and low, rich and poor, cultured and uncultured, sophisticated and simple; to the countless generations of the past as well as to the men and women of the future. I need a spiritual bond that binds me to all human beings in a common understanding of that which is fundamental and essential to human life. To have my life and happiness in that which is inaccessible to the many or to the few, seems to me an act of treason to humanity, a cowardly and pusillanimous attack upon the brotherhood of man; for without the inner spiritual tie of an essential aim which all can reach and all can understand, the concept of the human race as a spiritual unity is destroyed, and nothing is left of mankind but a biological species, only slightly better equipped than the other animals to cope with the present state of their physical environment. The differences between man and man are indeed inseparable from this our imperfect temporal existence; but I cannot and will not believe that their development constitutes the perfection of life itself. Rather is this to be found in the discovery and expectation of something underlying and absolute, something that can be found by all who seek it in earnest, something to which our lives may give expression, subordinating to its unifying principle the infinite multitude of ends, reducing them to their own relative measure and proportion, and refusing to permit the unimportant to become important, the relative to become absolute. The possibility of making this discovery and of giving it expression is, so it seems to me, the fundamental meaning of life, the source of its dignity and worth. The happiness that is found with this discovery is not invidious and divisive, but unifying and reconciling; it does not abrogate the differences, but it destroys their power to wound and to harm, the fortunate through being puffed up in arrogance and pride, the unfortunate through being depressed in envy and disappointment. For this happiness is not denied to any man, no matter how insignificant and humble.

Our criticism has brought us to the threshold of an ethical view of life. That the essence of life and its happiness is to be sought in the moral consciousness alone is the conviction that animates this address, and gives it its reason for being. This view holds that the individual human self has an infinite worth, that the personality has an eternal validity, that the bringing of this validity to expression in the manifold relations and complications of life is the true task of the self, that this task gives to the individual's historical development an infinite significance, because it is a process through which the personality in its truth and depth comes to its own. "Find your self," says the moral consciousness; "reclaim it in its total and in so far unworthy submergence in relative ends; dare to think the nothingness, the hollowness, the relativity, the precariousness, the lack of intrinsic meaning of that which constitutes the entire realm of the external and the manifold; liberate yourself from slavery to finite ends; have the courage to substitute the one thing needful for the many things wished for, and perhaps desirable, making first things first, and all other things secondary—and you will find that these other things will be added unto you in the measure in which you require them and can use them as servants and ministers of your highest good."

So speaks the voice within us, a still small voice, a soft whisper easily overwhelmed by the noise and traffic of life, but a voice, nevertheless, which no one can permit to be silenced except at the cost of acquiring restlessness instead of peace, anxiety instead of trust and confidence, a distracted spirit instead of harmony with one's self. The moral spirit finds the meaning of life in choice. It finds it in that which proceeds from man and remains with him as his inner essence rather than in the accidents of circumstance and turns of external fortune. The individual has his end in himself. He is no mere instrument in the service of something external, nor is he the slave of some powerful master; nor of a class, a group, or party; nor of the state or nation; nor even of humanity itself, as an abstraction solely external to the individual. Essentially and absolutely he is an end; only accidentally and relatively is he a means. And this is true of the meanest wage slave, so called, in industry's impersonal machine—precisely as true of him as it is of the greatest genius or the most powerful ruler.

Is there anyone so little stout-hearted, so effeminately tender, so extravagantly in love with an illusory and arbitrary freedom, as to feel that the glorious promise of such a view of life is ruined, its majestic grandeur shriveled into cramped pettiness, because the task which it offers the individual is not only an invitation, but also an obligation as well? Shall we confess that we cannot endure this "Thou must" spoken to ourselves,[1] even when the voice proceeds from no external power but from our inmost self, there where the human strikes its roots into the divine? Truly, it is this "Thou must" that is the eternal guarantee of our calling, the savior of our hope, the inspirer of our energy, the preserver of our aim against the shiftings of feeling and the vicissitudes of circumstance. It steels the will and makes it fast; it gives courage to begin after failure; it is the triumph over despondency and despair. For duty is the eternal in a man, or that by which he lays hold of the eternal; and only through the eternal can a man become a conqueror of the life of time. It is in the moral consciousness that a man begins truly to sense the presence of God; and every religion that has omitted the ethical is in so far a misunderstanding of religion, reducing it to myth and poetry, having significance only for the imagination, but not for the whole nature of man as concrete reality. The moral consciousness is a lamp, a wonderful lamp; but not like the famous lamp of Aladdin,[2] which when rubbed had the power to sum-

---

1. Suggested by Emerson's "So nigh is grandeur to our dust." *Voluntaries.*
2. S. Kierkegaard, *Postscript,* p. 124.

mon a spirit, a willing servant ready and able to fulfill every wish. But whenever a human being rubs the lamp of his moral consciousness with moral passion, a Spirit does appear. This Spirit is God, and the Spirit is master and lord, and man becomes his servant. But this service is man's true freedom, for a derivative spirit like man, who certainly has not made himself, or given himself his own powers, cannot in truth impose upon himself the law of his own being. It is in the "Thou must" of God and man's "I can" that the divine image of God in human life is contained, to which an ancient book refers when it asserts that God made man in his own image. That is the inner glory, the spiritual garb of man, which transcends the wonderful raiment with which the Author of the universe has clothed the lilies of the field, raiment which in its turn puts to shame the royal purple of Solomon. The lilies of the field[3] cannot hear the voice of duty or obey its call; hence they cannot bring their will into harmony with the divine will. In the capacity to do this lies man's unique distinction among all creatures; here is his self, his independence, his glory and his crown.

I know that all men do not share this conviction. Youth is often too sure of its future. The imagination paints the vision of success and fortune in the rosiest tints; the sufferings and disappointments of which one hears are for youth but the exception that proves the rule; the instinctive and blind faith of youth is in the relative happiness of some form of external success. Maturity, on the other hand, has often learned to be content with scraps and fragments, wretched crumbs saved out of the disasters on which its early hopes suffered shipwreck. Youth pursues an ideal that is illusory; age has learned, O wretched wisdom! to do without an ideal altogether. But the ideal is there, implanted in the heart and mind of man by his Maker, and no mirages of happiness or clouds of disappointment, not the stupor of habit or the frivolity of thoughtlessness, can entirely erase the sense of it from the depths of the soul. The present generation of men— particularly in the circles of the cultured and the sophisticated, those who are often called "intellectuals" and who perhaps also think of themselves as constituting a special class, characterized by a particularly acute awareness of life—the present generation exhibits in marked degree a loss of faith and enthusiasm, which it is pleased to call "disillusionment," and which perhaps also is disillusionment. They think of themselves, these moderns, as beset with despair;[4] the values that formerly seemed to be unquestionable have somehow gone dead, and many of them find nothing by which they are enabled to see life as dignified and serious. By and

3. S. Kierkegaard, *The Gospel of Suffering*, pp. 174–77.
4. *Postscript*, pp. 327, 328.

large, this despair is an aesthetic despair, an imperfect despair, which has not yet reached the ethical, or grasped the boundless meaning of the moral realm in its truth. Morality is for them not an infinite spontaneity, an inner life, an emancipation and an ennoblement of the self; it is for them mainly a system of conventions and traditional rules, an arbitrary burden imposed from without by social forces that have been outlived; or it is a mere device for reaching finite ends, whose worth has become doubtful. In so far as this despair is an aesthetic despair, it is all to the good; for this is the road which the spirit of man must take in order to find itself. Let us but learn to perceive that no differential talent, no privileged status, no fortunate eventuality, can at bottom be worth while as a consummation; that all such things are quite incapable of dignifying life; and when the misunderstandings with respect to the nature of a moral consciousness have been cleared away, the road will be open to the discovery of man as man. A preoccupation with the secondary thoughts and interests of life is always exhausting and trivializing, and in the end bewildering. Our true refreshment and invigoration will come through going back to the first and simplest thoughts, the primary and indispensable interests. We have too long lost ourselves in anxious considerations of what it may mean to be a shoemaker or a philosopher, a poet or a millionaire; in order to find ourselves, it is needful that we concentrate our energies upon the infinitely significant problem of what it means simply to be a man, without any transiently qualifying adjectives. When Frederick the Great asked his Court preacher if he knew anything about the future life, the preacher answered, "Yes, Your Majesty, it is absolutely certain that in the future life Your Highness will not be king of Prussia." And so it is; we were men before we became whatever of relative value we became in life, and we shall doubtless be human beings long after what we thus became or acquired will have lost its significance for us. On the stage some actors have roles in which they are royal and important personages; others are simple folk, beggars, workingmen, and the like. But when the play is over and the curtain is rolled down, the actors cast aside their disguises, the differences vanish, and all are once more simply actors. So, when the play of life is over, and the curtain is rolled down upon the scene, the differences and relativities which have disguised the men and women who have taken part will vanish, and all will be simply human beings. But there is this difference between the actors of the stage and the actors of life. On the stage it is imperative that the illusion be maintained to the highest degree possible; an actor who plays the role of king as if he was an actor, or who too often reminds us that he is assuming a role, is precisely a poor actor. But on the stage of life, the reverse is the case. There it is the task, not to preserve, but to expose, the illusion; to win free from it while still retaining one's disguise. The disguising garment ought to flutter loosely about us, so loosely that the least wind of human feeling that blows may reveal the royal purple of humanity beneath. This revelation is the moral task; the moral consciousness is the consciousness of the dignity that invests human life when the personality has discovered itself, and is happy in the will to be itself.

Such is the view of life to which the present speaker is committed. He has sought to make it seem inviting, but not for a moment has he wished to deny that it sets a difficult task for him who would express it in the daily intercourse of life. Perhaps it has long since captured our imaginations; for it is no new gospel worked out to satisfy the imaginary requirements of the most recent fashions in human desire and feeling; on the contrary, it is an old, old view. But it is not enough that the truth of the significance inherent in having such a view of life should be grasped by the imagination, or by the elevated mood of a solemn hour; only the heart's profound movement, the will's decisive commitment,[5] can make that which is truth in general also a truth for me.

5. *Postscript,* p. 226.

⊰ 3 ⊱

DAVID F. SWENSON

The Transforming Power
of Otherworldliness

The modern spirit exerts two powerful pressures on the religious address. On the theoretical side it insists on placing in the foreground of attention the problem of assimilating modern knowledge. The religious address is tempted either to defend religion against a possible attack from this quarter or to seek support for religious ideas in the results of recent scientific research or in recent philosophical speculation. On the practical side there is a tendency to seek a justification for religion in the eyes of common sense, to enroll religion in the service of the wisdom of experience, and to show how valuable it is as an instrument of efficiency. The promise of godliness for the life to come retreats into the background; its promise for the life that now is monopolizes the attention; and—what is more significant still—the latter promise tends more and more to be so interpreted as to bring it into complete accord with that natural desire for success which is so deep-rooted in human beings.

These influences exert their pressure upon the present speaker as upon others. Nor have I any wish to condemn them unqualifiedly as temptations to be shunned or evils to be resisted. The search for intellectual clarity has its obvious and genuine significance, and much that may be said about the contributions of religion to efficiency and energy in the business of life seems to me just and sound. But the tendency to an exclusive preoccupation with such considerations is another matter. This cannot but tend to obscure the distinctiveness of the religious message, and the qualitative heterogeneity of the religious principles.

Religion claims a place in human life, but the place it claims is not one of tolerance, but of supremacy. True religion does not borrow significance from other forms of human culture; it does not ask to be supported

31

either by science or by art, although it makes free to utilize both for its own distinctive ends. It does not wish to be impressed into the service of a kind of well-being or happiness external to its own spirit; it seeks to vivify and transform these conceptions themselves, and thus to be the arbiter of human valuations. It is, in brief, the master passion. It seeks control of "all thoughts, all passions, all delights, whatever stirs this mortal frame"; it strives to make all human activities its ministers—servants that feed the sacred flame of faith.

To speak otherwise of religion is to rob it of its spirit of aggressiveness. But the aggressiveness of religion is so much its very heart and soul that it cannot be defended or patronized without suffering betrayal. And yet it happens, strangely enough, that its representatives sometimes appear in the role of insinuating beggars, seeking the crumbs that fall from the table of the scientist or the philosopher or the worldly-wise man. In its truth, however, the religious spirit is a bestowing spirit, capable of making all rich, both rich and poor.

In this conviction I have chosen to speak this evening on the power exerted in human life by the spirit of otherworldliness. I do not mean to deny the significance of this-worldliness, or that religion has a this-worldly aspect. On the contrary, I claim for religion an inclusiveness, a suppleness, a quality of penetration, an absolute commensurability, which makes it at home in every situation. But it is my conviction that these two aspects of life, this-worldliness on the one side and otherworldliness on the other, are in their truth to be valued as the relative is to the absolute, the subordinate to the dominant, the many things useful to the one thing needful. Take away from life every relation to the absolute, and the relative itself is puffed up into the semblance of an importance that it cannot sustain. Take away from life every relation to the absolute, and the relativities of life lose their adequate ordering principle; life tends to become an anarchic chaos, or "a tale . . . full of sound and fury, signifying nothing."

Before proceeding with this talk, I wish to acknowledge my indebtedness to one of the great religious thinkers of all time, the Danish philosopher Søren Kierkegaard, whose name, although he has been dead three quarters of a century, is still practically unknown to English-speaking people, nor have any of his numerous works as yet found their way into English.* Because he created a new religious philosophy which stressed the category of the individual and his relation to God, and showed that truth can exist only in subjectivity, I feel that he is destined to exert a powerful ethical and religious influence upon future thinking, and I am therefore glad to afford you a brief introduction to some phases of his thought by drawing freely upon the ideas he set forth in one of his edifying discourses, entitled, "The Expectation of an Eternal Happiness," published in 1844;

* This was true when Swenson wrote this essay; it no longer is. [Ed.]

and by paraphrasing certain passages I shall try to give you some idea of the clarity, the simplicity, and the earnestness with which Kierkegaard appealed "to that individual" whom he was happy to call "my reader." Hence, employing the same text as did Kierkegaard, I call your attention to certain moving words of the Apostle Paul, which you will find in his Second Letter to the Corinthians, words that express the conviction embodied in my theme:

> Wherefore we faint not; but though our outward man is decaying, yet our inward man is renewed day by day. For our light affliction, which is for the moment, worketh for us more and more exceedingly an eternal weight of glory; while we look not at the things which are seen, but at the things which are not seen: for the things which are seen are temporal; but the things which are not seen are eternal.

These words of Paul testify to a conviction which his life gloriously interprets. The principle of this conviction is the thoroughgoing interpenetration of life with the consciousness of the eternal, the bringing of the invisible everywhere to bear upon the visible, finding in the eternal the ultimate ground and the final clarification of human life.

The modern spirit is not indisposed to admire the boldness of Paul's faith, but it hesitates to follow in his footsteps. Its reasons are multifarious, but chief among them is the lurking fear that such faith is, after all, a retreat from life into the realm of fancy. It must be admitted that we are all of us confronted with this danger, a danger that presents itself not in one form only but in many, not only when life is professedly religious, but also when life is thoroughly secularized. Imaginary compensations for actual weakness, the selfish and cowardly indulgence in daydreams that but increase that weakness more and more, sapping the soul's vitality—these are a sickness of the soul which may threaten to become a sickness unto death. The spirit of otherworldliness in religion is not only intolerant of this weakness but is an actual cure for it. The hope of eternal bliss is for Paul not a refuge from life; it is a challenge to live yet more deeply and intensely. The essence of daydreaming is the separation of the imagination from the will, its development in isolation from the rest of life. The essence of a religious faith in an external order is its transforming power over the rest of life, an influence that dares ignore no single detail as too small or unimportant. The eternal presses instantly upon the will; its motto is: "Work . . . while it is day"; "Now is the acceptable time." If the expectation of the eternal does not work this result upon the life, it is a counterfeit expectation and not a genuine faith, a credulous superstition, an unclarified yearning, an affectation, an insincerity, a sick soul's flight from life, not a sound soul's sincere presence in the temporal.

Those who do not share the enthusiasm of this faith naturally regard it as an illusion. Nevertheless, it must not be forgotten that faith in the eternal arises like the phoenix out of the ashes of all the soul's previous illusions. It supersedes not only the happy illusions of childhood and the beautiful illusions of youth, the illusion of a romantic love that is happiness without responsibility; the illusions of maturity, which center about fame and money, power and success; but also the most persistent and deep-seated of all illusions, the illusion of self-adequacy and self-righteousness. This supersession is that dying away from the world of which religion speaks, a dying that conditions the life that is life indeed; the pain of this dying is mingled with the joy of this living; and this most significant of all the struggles that life affords is the supreme test of the reality of the soul's thoughts and aims and feelings.

Human life sometimes affords us a glimpse of an aesthetic perfection from which this deeper struggle seems to be excluded. The life of creative art, for example, appears to have a perfection of unity that knows nothing of the deep cleft between preparation and consummation. Each step or phase of such an activity seems both an end in itself, yielding its own immanent satisfaction, as well as a transition to the equally significant next step or stage. But such experiences are fortunate episodes in life, not expressive of its essential structure. They are imperfect analogues, foreshadowing dimly that deeper harmony which can enter into the present life only by anticipation and faith. Man is a duality;[1] such is the testimony of one of the greatest thinkers of antiquity, whose mythical genealogy of Eros as the child of Poverty and Plenty, who because of his participation in this double nature is always striving, desiring, seeking for that which in its fullness is beyond the present self, is an imaginative fancy meant as an essential picture of human life.

Such is also the testimony of Paul, who makes human life a synthesis of the temporal and the eternal, faith the anticipation that knits them together, eternity the true fulfillment. Time separates the way from the goal to which it leads; it separates means from ends, preparation from consummation. Every more earnest view of life acknowledges the significance of this separation; only frivolity expects to reap its harvest the instant it has sown. The more ambitious the view of life, the more remote the consummation that it proposes; this principle is, in general, the measure both of the worth of the striving and of the significance of the reward sought. The religious view of life is the most ambitious of all; it is charged with a demand upon life that makes even the most beautiful temporal consummation seem a sorry imperfection, if it is to be *the* consummation. It dis-

1. Plato, *Symposium*; S. Kierkegaard, *Postscript*, p. 54.

poses of an enthusiasm that is content to make, not a longer or shorter period within time, but the entire earthly life a state of striving and of preparation. The religious man confesses to the view that life is a way, that a man is, in his true character, a pilgrim; he looks toward an unseen city, "a house not made with hands, eternal, in the heavens." In testimony of this he calls himself a believer, which means in his mouth what the staff signifies in the wayfarer's hands, namely, "What I seek is not yet here."

Let us now consider this view of life from a variety of aspects in order that we may be enabled to grasp more clearly the part that it plays in human life; the difference that is made, the power that is exerted, by the expectation of an eternal life and an eternal joy.

First, let us note that this expectancy gives to our lives their greatest possible tension, since it evokes the maximum interest. To bestow an infinite interest upon a finite value is the torturing self-contradiction of the mere worldly life. It is not only hope that is found to be precarious, but experience teaches us that the things hoped for deceive us even though they be attained. The fulfillment at hand, we ask ourselves, "Is *this* the consummation I so devoutly wished?" The eternal alone can furnish the true objective for that infinite interest which God has implanted as a potentiality in the heart of man. And, conversely, it is the evocation of this interest which constitutes the core of the true belief in immortality. Where this passion is atrophied or suppressed, the idea of immortality is a mere luxury, an unmeaning decorative addition to the surface of life, an affectation and a fancy. It is this deep and vital interest that is the oil in the lamps of all those who wait the coming of the bridegroom, the neglect of which excludes one from participation in the marriage feast; for an infinite personal interest is the subjective form for an indwelling of the eternal. When once the possibility of such an interest has dawned upon man's horizon, he is confronted with a challenge which he cannot evade or deny without doing violence to his deeper self; this glimpse of immortality makes a belief in it morally necessary, in the same sense that a philosopher once said of God that if God is possible,[2] then he is also necessary.

Let us also note that the consciousness of immortality removes the last vestige of abstractness, of casualness, of the possibility of subterfuge and evasion from man's relationship to God, and reveals life in its absolute earnestness. A transitory relation has in it the seeds of unreality. God's immanence in the world and in the human consciousness is an invisible presence; it does not have the immediately obvious or compelling character of an external or sensible reality. His omnipresence is therefore

2. Voltaire.

a seeking, a questioning, a solicitation—a courting, if you will, of man's free spirit; and this unobtrusive courtship is the most tremendous concession conceivable to man's independence. But lest this concession become a mere weakness on the part of God, a sheer madness, it must be permeated by the condition of an inescapable responsibility on the part of man. But if, after a lifetime of evasion and neglect, I can end all by simply tucking myself away in a grave, and be as if neither I nor God had ever been, then God has permitted himself to be mocked, and life is not real and earnest to the last degree. But God is not mocked, and it is for this reason also that he is not the God of the dead, but of the living. The true faith in immortality is thus nurtured in fear and trembling.

The expectation of immortality presents to the individual an ideal goal in which he can be concretely present, because it is a goal in whose attainment he himself participates. There are views of life that seek to enhance its ideal significance by setting the individual a task whose realization is placed beyond the individual life, in some future period of mankind or of the world. This view makes all the preceding generations of men mere instruments, not ends in themselves; it arbitrarily selects a particular generation, or a group of generations, to be the fortunate beneficiaries of an effort which they have not shared, and permits them to reap where they have not sown. The concept of the progress of the race doubtless has its relative meaning and justification; but alone and by itself as the highest good, it is a mere abstraction. If, on the other hand, every human being has his ideal goal in himself, and if this self is no mere particularistic entity but a true individual, at one and the same time itself and the human race, so that men cannot without humanity be made perfect, then there is established an immanent end for man, which makes him no mere instrument, and makes of his life something higher than slavery in the service of external ends. And the human race too becomes something more than an intellectual abstraction, as it is in the historical process; in the relationship of each and every man to the eternal, the human race as such becomes a concrete reality.

The expectation of immortality has the power to reconcile every human being to his neighbor, his friend, and his foe, in a common understanding of the essential in life. It unites him to his fellow men in the most profound of all sympathies, and in the only thinkable universal goal. Every other goal is to a greater or less extent divisive, depends upon differences, and establishes invidious distinctions, not forgetting the invidious distinction between those who are privileged to give and those who are foredoomed merely to receive. But the expectation of immortality is not relevant to differential talent or fortunate circumstances; it establishes itself upon the basis of our common humanity, and no one is excluded who does not exclude himself. It cultivates no stoic insensibility; it acknowledges the genuineness of relative goods, but it denies that they constitute the perfection of life. It looks, therefore, toward a more perfect life, in which all men will have all things; in which God, who is all in all, will by the miracle of his love be present whole and entire in every single human heart. To believe something like this, so it seems to me, is to be human. And believing this, one may learn to wear his differences lightly, as if they were but the disguising cloak for the royal purple of humanity beneath.

The expectation of an eternal issue for human life will also help to liberate a man from the ruinous illusion that the quality of the means he chooses is of less importance than the ends he seeks. It is of the essence of the eternal, as the goal of life, that method and end are always in exact and perfect harmony. The methods, on the contrary, by which any temporal end may be attained, are diverse and incommensurable, and the choice of means becomes more or less a matter of indifference. He who stakes everything upon his success in attaining such ends will in the last analysis succumb to necessity that knows no law. Every ethical principle will tend to become for him vacillating and uncertain. Honesty is the best policy—aye, in most circumstances perhaps—at any rate the appearance of honesty, if not the reality, and a certain degree of honesty, if not its utmost perfection. And so with every other ethical principle; it must not be applied too severely, lest it defeat its own ends and fail to be justified by the results. The maxim of a worldly prudence is therefore always *ne quid nimis,* to carry nothing to extremes. It has no sympathy with that divine madness, the transcendent enthusiasm, which breaks through the confines of a petty calculation, because it derives its strength from the eternal and the unchanging. The statesman permits himself an appeal to the lower levels of character and intelligence, and thus abdicates his leadership before he enters upon it, because he fears that any other course would jeopardize an immediate success. All who make it their highest aim to secure for themselves any form of external power—kings and emperors and statesmen and demagogues and ecclesiastics and sophists—have not scrupled to exploit and foster human weakness and illusions, its prejudices, and its basest passions. They boast of taking the world as they find it; it cannot escape a thoughtful contemplation that they also leave the world in essentially the same situation. Not so the saints, the apostles, the prophets, and the martyrs, the men and women who have yielded allegiance to an eternal order, and have dedicated their lives to bear witness for the truth. Concession and distortion might have brought them a tremendous success; but they lived for eternity, and hence chose rather the threat of failure and the certainty of opposition. And eternity knows them, and acknowledges them for its own. In the perspective of the moment, success is everything, and failure is failure; in the longer perspective of history, the immediate success is already of lesser importance, and the failure of the moment is sometimes the success of the future; in the perspective of eternity, success is nothing and failure is nothing, but the spirit of endurance in suffering and loyalty in striving for a good cause is everything. The expectation of the eternal will help a man to be steadfast and true to his cause; it will help him to understand the difference between the devotion that serves a cause and the calculation that attaches itself to a cause in order to be served by it; and it will habituate in him the search for the most glorious harmony of human life, the harmony of the means with their corresponding end.

The expectation of the eternal will help a man to be contemporary with himself in time, and wean him from the paralyzing effort of trying to shoulder the vaguely looming burden of tomorrow, or of all the tomorrows, in place of a concentration of effort and attention upon the definitely restricted burden of today. The power to concern himself with the future is a mark of man's kinship with the divine. For if there were no future, there would be no past, and if there were no past and no future, man would be hidebound in his instincts and a child of the moment, his horizon no broader and his head no higher than that of the beast that perishes. The struggle with the future is therefore an ennobling struggle, and no man can conquer the present who has not first conquered the future. But how can a man conquer the future before he has fought with the present? Only through that expectation of victory which is of the essence of faith in the eternal. This expectation is no mere youthful buoyancy, no mere confident self-appraisal. It is an expectation of a victory that transcends both success and defeat, both happiness and sorrow. It is the assurance that all things work together for good to them that love God. When the future has been conquered in this sense and in this sign, then and only then can the individual soberly and without illusion be wholly present in the present task, in each successive moment contemporary with himself.

Finally, let us note that the expectation of an eternal happiness gives a man a measuring rod, a standard of estimation, which will help him to understand himself in time, as over against the extremes of stress and strain, of fortune and misfortune. The wisdom of experience yields some such standard, and is useful in estimating the individual moments of life, mediating an adjustment to them that lends to man some degree of poise and equanimity. But this wisdom is based upon the individual and the particular, upon the partial and the discrete; the measure that it yields is a restricted and finite measure, the application of its standard confined within a limited compass and sphere. Whenever a man's soul is pregnant with the expectation of the eternal, this measure becomes for him too small and petty; he may thus, unwisely, be tempted to reject its relative aid. But neither can he fall back upon the romantic standards of youth. Whoever has only a finite standard of evaluations has a life that is wholly a child of time, and cannot even be assured of his ability to cope with the finite, since it is always possible that events will break that his experience cannot help him to surmount. If this happens, he becomes the prey of despair. Whoever has experienced this has learned from experience that experience is not enough. But he who cherishes in his soul the lively expectation of an eternal joy has in that expectation a measure that is always adequate and always valid. By means of this standard he will be enabled to understand himself in times of stress and strain, when life's demands exceed the measure of a finite understanding, and the wisdom of experience breaks under a pressure it cannot bear. He will not permit success or happiness to rob him of this eternally valid standard, and substitute in its place the false and transient standards of vanity and conceit; he will not permit sorrow and suffering to impose upon him their false standards, and teach him the despairing grief that knows no hope.

"For our light affliction, which is for the moment, worketh for us more and more exceedingly an eternal weight of glory." Let us imagine for a moment that we know nothing more about the author of these words than the fact that he had said them. And let us assume that we tried, as well as we knew how, to infer from these words what sort of life must have fallen to his lot—to the lot of one who could give this earthly life and the life to come such a testimony. Doubtless we should conclude, should we not, that this man's life must have been relatively quiet and uneventful, a life in honorable obscurity, far from the turmoil and stress of great crises, not perhaps entirely unaware of that secret wisdom which suffering bestows upon its initiates, but nevertheless not tried in the extremes of physical danger and mental anguish. We should perhaps be reminded of another saying of apparently similar purport and background, namely, "that the earth is beautiful enough as a country to travel through, but not so beautiful as to make the wayfarer forget that he is on his way." And, having drawn this or a similar conclusion, let us suppose that someone then outlines for us, as if for the first time, the story of Paul's life. He tells us of Paul's spiritual trial on the road to Damascus, when he was made to reverse his whole concept of right and wrong, and hence evidently was a man not without experience of the soul's supreme crisis; he tells us of Paul's mystical exaltation and rapture, his being lifted up into the third heaven—and so we see that Paul was doubtless tempted to conceive a distaste for the ordinary humdrum course of life; he tells us that Paul was often heard to testify with an enthusiasm that made him seem a madman; that for the many years of an active life he was without home and without security, a scandal in the eyes of his compatriots, a fool to the Greeks, an outcast from the world, often in danger to life, experienced in hunger and nakedness, often in prison, and at last executed as a criminal.

Would not this narration tend to shake us out of ourselves, our poise and our equanimity, and extort from our lips the exclamation that such a life was not to be endured—an everlasting strain and the heaviest affliction? The standards of our experience would break down under the attempt to compass and to estimate such a life. But Paul had an eternal goal, and an eternally valid measure; hence it is that where we speak of a heavy and unendurable burden, he spoke of "a light affliction, which is for the moment." When the storms of life threatened to overwhelm him, Paul looked toward his goal, and measured the storms with his standard, and behold, they became light and momentary afflictions! When experience failed to help him, when its wisdom fell short of the demands of life, then his faith leaped to the rescue, laying hold of a joy and peace beyond all understanding.

When the stress of life transcends the measure of experience, and plunges worldly wisdom into the depths of despair, human life becomes a chaotic turmoil, without comfort and without hope, unless the hope of an eternal weight of glory brings into our lives its order and its calm. This is that genuine transvaluation of all values which religion through its otherworldliness brings to life: not the fantastic transvaluation of an imaginary superman, but the simple faith of common men and women, a faith, however, from which no genius need be excluded unless he excludes himself.

≫4≪

REINHOLD NIEBUHR

The Self
and Its Search
for Ultimate Meaning

In analyzing the various dimensions of human selfhood, we have thus far concerned ourselves with the presuppositions of the inquiry only insofar as we have called attention to the fact that the tendency to identify the self with its mind is as erroneous as it is persistent. The error obscures the freedom of the self over its rational faculties.

If we now proceed to inquire more rigorously into the dimension and character of that freedom, it will become apparent that the religious inclination of men is derived from that freedom. The freedom makes it impossible for them to consider systems of rational intelligibility, whether conceived in idealistic or naturalistic terms, as a solution for the problem of the meaning of their life. They discern a mystery and meaning above and beyond their rational faculties in themselves; and they also surmise that the chain of causes, whether conceived in terms of efficient or final cause, that is, whether in terms of idealistic or naturalistic metaphysics, points beyond itself to a mystery of creativity. This is true because any previous event is an intelligible, but not a sufficient, cause for the succeeding event. It was Hume's scepticism about causes which first breached the confidence in the intelligibility of the world and prompted Kant to elaborate his critical idealism, in which for the first time the confidence in reason was separated from confidence in a rational ontological order. The result was his conception of a mysterious *Ding an sich,* of a noumenal reality which indicated the mystery beyond the rationally intelligible phenomenal world of finite entities and relationships.

These scruples in modern philosophy have tended to create mystic overtones in any rational system, well-defined in Bertrand Russell's *Mysticism and Logic*. They are indications of the inevitable emergence of explicit religion even in cultures which have ostensibly banished mystery from a world believed to be completely subject to a rational order including both the knower and the known.

It must be noted however that the world views which assume the rational intelligibility of the world, without mystery, are not less "religious" because they disavow explicit religious faith. Confidence in a rational order is, as the great rationalist Bradley admits, also a faith. The self, even of a philosopher, is religious to the degree that the self must commit itself to a system of meaning, even if it has the view that the system is so self-explanatory and that it takes the self as mind so completely into its system of self-consistent truth that nothing more than the elimination of ignorance seems required to prompt the self to its acceptance.

But it is significant that these implicit religions never suffice in the long run, even when, as in the French Enlightenment or in modern communism, they rise to emotional heights of devotion which inadvertently betray a committed self rather than a dispassionate mind. The more explicit religions are, however, the more overt reactions to the fact that the self senses a mystery in itself and a mystery in the world beyond the flux of observable causes. It therefore tries to overcome the threat to the meaning of its life by finding that the one mystery, the ultimate or divine mystery, is a key to the understanding of the mystery of the self's transcendent freedom.

The task of penetrating the ultimate mystery prompts many responses, but they could all be placed into three general categories: (A) The first category embraces all religious responses in which the self seeks to break through a universal rational system in order to assert its significance ultimately. It may seek to do this individually, as in modern romantic and existentialist thought; or it may be so conscious of its finiteness as an individual that it finds no opportunity to assert the ultimate significance of itself in history except by asserting the significance of the collective self. This category, in short, embraces all the idolatrous religions of ancient history, including both primitive polytheism and the imperial religions of Egypt and Babylon, and (in more artificial terms) of Rome. Until a recent day this idolatry, in which the individual self finds the ultimate source of its meaning in the history of the collective self so much more imposing though also so much closer to the flux of nature, was thought to be a phase of history which was overcome by the rise of the rigorously monotheistic religions and monistic philosophies. But the recrudescence of religious nationalism and the pseudo-universalistic Messianism of communism have instructed us that this idolatry, this worship of the collective self as if it were ultimate and not finite, is not merely due to the limits of a primitive imagination. It corresponds to a perennial desire in the human heart to eat one's cake and have it, too; to subordinate the finite self to something greater than it but not so great that the self may not participate in the exaltation of the finite value. Naturally this idolatrous religion must have baneful effects, not only because it complicates the problem of group relations by exaggerating the claims of contingent historical forces in competition with each other, but because the unconditioned commitment of the self to the collective self must rob it of its freedom; for the collective self is, though more imposing and more long-lived than the individual self, also so much more bound to nature and its necessities, so defective in organs of self-transcendence and therefore so much farther removed from the ultimate source of meaning, that the self debases itself by this uncritical devotion.

(B) The second alternative of explicit religious response has been defined by Aldous Huxley as "The Perennial Philosophy." He is right in asserting that it is a fairly universal response, but wrong in concluding that this universality guarantees its validity. This response, generally defined as "mysticism," stands at the opposite pole of idolatry. It is in fact an heroic effort to transcend all finite values and systems of meaning, including the self as particular existence, and to arrive at universality and "unconditioned" being. The persistence of this mystic tendency in the religions of the world is a telling proof of the ability of the self, in the ultimate reaches of its freedom and self-awareness, to discern some affinity between the mystery within itself and the mystery behind the observable phenomena and to find the key to universality in the joining of these two mysteries. This "perennial philosophy" embraces not only the systems, stemming from the thought of Plotinus, in the Western world but practically all religions of the Orient. It is expressed in the Brahman overtones of Hindu polytheism; in the Sufist tradition of Mohammedanism; in the Taoist tradition of Chinese culture and, most classically, in Buddhism. Here the search for undifferentiated being reaches the height of asserting a type of being as the goal of existence about which one can not be certain whether it is the fullness or the absence of being. It is certainly being bereft of all relationships and meanings.

(C) The third alternative, an explicitly religious answer to the self's search for the ultimate, embraces the two Biblical faiths of Judaism and Christianity. These faiths interpret the self's experience with the ultimate in the final reaches of its self-awareness as a dialogue with God. The idea of a dialogue between the self and God assumes the personality of God, an assumption which both rationalists and mystics find untenable, but to which Biblical faith clings stubbornly. Selfhood or personality is supposedly not attributable to God because the idea of personality is loaded with connotations of finiteness and therefore casts a suspicion of "anthropomorphism" upon Biblical faith. But it is significant that both mystics and rationalists have as much difficulty in ascribing personality to man as to God. This fact suggests that it is not the connotations of finiteness which create the difficulty but rather the fact that personality is characterized by both a basic structure and a freedom beyond structure. The rationalists can comprehend the structure within a system of rational cohesion; and the mystics are able to interpret the freedom as part of a system of undifferentiated potentiality. But neither is able to comprehend the total fact of personality within its system.

The dialogue between the self and God results in the conviction of the self, but not for reason of its finiteness. It is convicted rather of its pretension or "sin"; of claiming too much for its finiteness, and for the virtue and wisdom, which it achieves in its finiteness. The idea of such an encounter therefore permits the Biblical faiths both to affirm the life of the self in history and to challenge its achievements in any particular instance. "Enter not into judgement with thy servant, for in thy sight is no man living justified," declares the Psalmist. Kierkegaard sums up this theme of Biblical religions with the affirmation that "Before God all men are in the wrong." The fact that the self is judged for every inclination which affronts God's "majesty" by pride or lust for power is the religious dimension of sin. The prophets are however equally conscious of the social dimension which is the inclination of the self to take advantage of its fellow men. This "injustice" is never speculatively defined, as in Greek philosophy, but rigorously defined by reactions to injustice in particular situations.

The "severity" of God's judgement is matched by the "goodness" of His mercy. In the dialogue between the individual and God, this validates itself as the indeterminate possibilities of self-realization and fulfillment of the self's potentialities once it has ceased to seek fulfillment of life from the standpoint of itself. The problem of how the mercy of God is related to His justice is a perpetual problem in the Old Testament. The new Biblical faith of Christianity enters into history with the affirmation that the drama of Christ's life is in fact a final revelation, in which this problem is clarified by the assurance that God takes the demand of His justice upon Himself through Christ's suffering love and therefore "God was in Christ reconciling the world unto Himself."

The dying and rising again of Christ is the key to the self's possibilities in history. All of life is given this norm for the realization of selfhood. "I am crucified with Christ," declares St. Paul, "nevertheless I live." This theme is in perfect harmony with the words attributed to Jesus in the Johannine Gospel: "Except a corn of wheat fall into the ground and die, it abideth alone: but if it die, it bringeth forth much fruit." (John 12:24)

Thus the encounter of the self with God is defined in Biblical faith in terms of a norm which has been set by an historical "revelation." And this revelation is an historical event or series of events which are not essentially miraculous (miracles such as the "virgin birth" are afterthoughts) but are events in history which are discerned by faith to have revelatory power into the ultimate mystery. Both Biblical religions are covenant faiths, which organize covenant communities upon the basis of a common commitment of faith in the divine significance of these events. We must postpone until later, then, a consideration of the relation of revelation to the drama of history. In this connection it is necessary to observe that the discernment of ultimate significance of an historic event makes the Biblical religions seem primitive and unsophisticated in the eyes of both rationalists and mystics, who look for the ultimate or "unconditioned" in either the permanent structures of existence or in an undifferentiated ground of being. They may fail to note, however, that the Biblical presupposition is the only one of the three alternatives which asserts a discontinuity between the self and God. This discontinuity makes explicit faith indispensable in the ultimate dialogue; but it also prevents the self either from usurping the place of the divine for itself or from imagining itself merged with the divine. If we test these three alternative solutions for the self's search for the ultimate by the two tests of consistency or coherence with other truth, and by conformity with established facts subject to empirical tests, it will soon become apparent that the religions which tend to the exaltation of finite values and centers of meaning are most easily ruled out, as indeed they have been ruled out in principle for centuries. The collective self may be momentarily imposing; but its mortality is obvious and the perils to the individual self by its pretensions of divinity are very great.

It is however very significant that a religious solution which has been ruled out in principle for centuries should have so much practical force in our day, both in the version of a religious nationalism and in a pseudo-universalistic Messianic creed. These contemporary ventures into idolatry are proof of the difficulty of containing the collective self within any more general scheme of validity than its own interests. They prove that an affirmation of historical meaning as we have it in Western civilization is almost inevitably attended by pretentious efforts to close the system of meaning prematurely with some cherished value of the self at the center of the system.

It is equally significant that modern culture has generated less plausible and dangerous forms of individualistic pretention, in which the freedom and the uniqueness of the individual is asserted in defiance of any systems of consistency or universal meaning. The romantic revolt of the nineteenth century culminated in Nietzsche's effort to achieve the affirmation of unique vitality of the individual and his transcendence over the flux of history, thus seeking to combine classical with Hebraic attitudes toward time and eternity.

It must be apparent that modern existentialism is but another version of this romantic revolt. It has obviously learned from Biblical faith about the unique freedom of the individual and the distinction between the self's reason and personality. It is however unable to make the venture of faith of Biblical religion and therefore ends in the quasi-idolatrous attitude of making the individual his own creator and end. "Thus there is no human nature," declares the French existentialist Sartre, "because there is no God to conceive it. Man simply is. Not that he is simply what he conceives himself to be. But he is what he wills. . . . He is what he wills to be after that leap toward existence."[1]

Heidegger's concern for "authentic being," for the affirmation of the uniquely human freedom against the necessities of nature and the inevitability of death, is distantly related to Nietzsche's defiance of death. It is in the same category of quasi-idolatry. It may not make the self into its own God but it asserts the uniqueness of the self without reference to its relations to the community or to any general value.

If we rule out the idolatrous and quasi-idolatrous, the individualistic and collectivistic forms of these idolatries, as valid answers to the self's quest for ultimate meaning even though we recognize that the popularity of such answers is not confined to past history but is an ever recurring phenomenon, we are left with the two alternatives of the Biblical faith and Mr. Huxley's "perennial philosophy" or the mystic answer to the problem.

The answer of Biblical faith embodies, as we have seen, several presuppositions and affirmations which the modern mind finds particularly difficult, not to say impossible: the personality of God; the definition of the relation between the self and God as a dialogue; and the determination of the form of that dialogue in terms of a previous historic "relevation," which is an event in past history, discerned by faith to give a key to the character and purpose of God and of His relationship to man. It is therefore understandable that when confronted with these two alternatives, sophisticated moderns who have become aware of a depth of selfhood which can not be comprehended within the limits of the self as a biological organism or the self as mind, are inclined to turn to the mystic alternative in preference to the Biblical one. It is even understandable that they

1. *Existentialism and Humanism,* p. 28.

should do this at the price of defying the very ethos of their own life-affirming and history-affirming culture and choose an alternative which annuls every partial and particular meaning including the particular self. This is understandable in the sense that it proves how powerful are the compulsions to comprehend reality in a self-consistent scheme and to leave the mystery beyond the system of rational intelligibility unsolved.

Thus Professor Stace uttered a cry of despair some years ago because he became aware that the world which modern science explicated had no place for the human self or for any of the values which the self holds dear. Subsequently he published his considered answer to this problem in his *Time and Eternity.* He had accomplished his escape from the naturalistic prison by embracing the "perennial philosophy" of Mr. Huxley and the oriental mystics. He defined religion as the search for "the impossible, the unattainable and the inconceivable." Professor Stace thus bears testimony to the capacity of the self to reach for the ultimate; but he is sceptical of any venture of faith in an ultimate which would purify and complete the particular meanings of history. He finds it more acceptable to assert the pure mystery of the divine. He is impressed by the fact that the mystic approach arrives at the conclusion that God is both the fullness and the absence of being. Reporting on the account of divinity in the mystic tradition, he records that "God is non-Being, nothingness, emptiness, the void, the abyss . . . God is the great silence, the great darkness . . . yet God is also in the language of the medieval mystics, the supreme reality, the 'ens realissimum.' " "This supreme God," he declares, "is contrasted by the mystics with the worthlessness of the world . . . the world then is worthless trash. This is seen by all men more dimly or more clearly, but it is seen by the mystics with absolute clarity."[2]

Professor Stace refers frequently to the Hindu desire to achieve unity of the self and God, to realize the assurance that "Brahman and Atman are one." This seems to him to be pure religion in comparison with the religions of the Bible with their appreciation of particular selfhood.

The impulse to annul the meaning of particular selfhood and the significance of the whole drama of existence is expressed even more significantly in the view of the eminent philosopher George Santayana. Despite his essential Platonism, and his consequent faith in a "realm of essences," Santayana makes it clear, in his *Platonism and the Spiritual Life,* that the final goal of religion must be to transcend even these ghostly structures of particular meaning. "At the risk of parting company with Dean Inge, or even with Plato," he declares, "the spiritual life is not a worship of values, whether found in things or hypostasized into super-

2. *Time and Eternity,* p. 126.

natural powers. It is the exact opposite. It is the disintoxication from their influence. . . . The great masters of the spiritual life are evidently not the Greeks, not even the Alexandrian Greeks, but the Indians and their disciples elsewhere in the East; and those Moslems, Jews and Christians who have surrendered precisely that early unregenerate claim to be enveloped in a protecting world." Santayana makes it quite clear that such a faith annuls all historic possibilities and responsibilities. "Obligations are moral," he writes, "they presuppose physical and social organisms and immanent spontaneous interests. . . . All values fall within the preview of ethics, which is a part of politics. Spirituality is the supreme good for those who are called to it, the few intellectuals who can be satisfied only by the impartial truth and by the self-annihilating contemplation of all being."[3]

It is rather revealing that Santayana reserves the mystic *summum bonum* for a few intellectuals. It reveals how aristocratic is the conception; and how closely mysticism is related to rationalism. From Aristotle to Santayana, mysticism is in fact the perennial overtone of rationalism. The drama of history is not comprehended in the categories of meaning supplied by either the rationalists or the mystics. In the one case the categories fail to comprehend the dramatic variety and the complex causal relations of history. In the other case the mystic conception of the fulfillment of meaning obviously results in the annulment of any particular meaning in history.

It will be regarded as futile by all pure "empiricists" to compare the Biblical and the mystic conceptions of the ultimate dimension of selfhood and to judge between the thesis that the self is in "dialogue with God" and the thesis that the self on that level is in the process of merging with a universal divine consciousness. But if the evidence of introspection is accepted (though it is admittedly inexact) it can not be too difficult to prove that the abstraction of the universal subject from the self as particular object is a futile procedure because the particular self always remains obtrusively in these exercises of introversion. There is furthermore the social evidence that the mystics never succeed in eliminating particular selfhood or in transcending the self as a particular organism. The erotic overtones in the mystic visions of an absolute consciousness is a rather pathetic symbol of the futility of the self's attempt to escape from the "body" and time into an undifferentiated eternity.

In short, we are confronted with evidence that the thesis of Biblical faith, that the self is in dialogue with a God who must be defined as a "person" because He embodies both the structure of being and a transcendent freedom, is more valid than the alternative theses which find

3. Pp. 34–40.

much greater favor among the sophisticated. The Biblical thesis requires a more explicit act of faith because it leaps a gap of discontinuity between man and God and because it dares to give a specific meaning to the divine, which is relevant to the partial and fragmentary meanings of history. It both fulfills and corrects these meanings, loyalties, and values, and therefore has a more valid attitude to the self's historic existence which the various rationalistic systems affirm too simply and the mystic thesis annuls too absolutely. This character of Biblical faith is therefore the crux of the question, why a faith which is more explicit than alternative ones should be more justified by actual experience than these. It gives a key to the seeming mystery of our whole cultural history. That mystery is why an allegedly "dogmatic" faith should be justified by the experiences of the human self more than the allegedly "empirical" approaches to selfhood, which obscure their potent, though implicit, dogmas within their prescriptions for empirical observation.

We must examine this strange paradox by an historical analysis of the long debate between the Biblical-dramatic and the rational approaches to the problems of the self in the history of Western civilization. . . .

A concluding word must be said about the intimate relation between the hope of the fulfillment of individual selfhood with the hope for the fulfillment of the whole historical drama. How incredible and how valid this combination of hopes is! It is valid precisely because the individual self is grounded in a collective history as surely as it is based in a physical organism. Its fulfillment is not possible without the fulfillment of the whole drama, yet the fulfillment of the total drama offers no adequate completion of meaning for the unique individual. One is reminded in this instance of the many humble souls of either pathetic or noble proportions. Historians may fit the lives of the thinkers and the statesmen of history into some kind of correlation which endows their lives with meaning. But what is to be done about the humble spirits who do not rise to the height to be fitted into these correlations and schemes of meaning? And for that matter what is to be done with these same statesmen and philosophers insofar as they are anxious individuals rather than great thinkers and doers? How, in other words, can we bring the whole human story, including all the relevant and irrelevant individual dramas, into some scheme of intelligibility without obscuring and denying the richness and variety of the drama? The hope of the forgiveness of sins and life everlasting is thus a fitting climax of the faith that there is a meaning to the story beyond our understanding of its meaning because it is grounded in a power and purpose beyond our comprehension, though not irrelevant to all our fragmentary meanings. It is the final venture in modesty for the mysterious human self, which understands itself more completely if it understands that there are heights and depths of human selfhood which are beyond any system of rational intelligibility, but not beyond the comprehension of faith and hope.

The question about interpreting human selfhood and its dramas within either a framework of rational intelligibility or a framework of meaning and mystery, which is the question between modern culture and faiths which are rooted in the Bible, can be briefly summarized. There are directly experienced realities in the realm of selfhood which point beyond any system of rational intelligibility to mystery, and they create a sense of meaning and mystery in which the one penetrates the other. Such realities or "facts" are the freedom and responsibility of the self, its sin and guilt, the unity of its freedom and its physical organism, and the variegated dramas and dialogues of which it is capable. These facts can not be contained in a system of rational intelligibility because there are no causal sequences and coherences which could contain the facts.

The systems proposed by classical or modern philosophers or scientists invariably deny or obscure some directly experienced facts for the sake of the coherence of the system. The facts point beyond themselves to a realm of mystery. But pure mystery destroys meaning as surely as pure intelligibility. The genius of Biblical faith is that it discerns by faith, glimpses of meaning in the ultimate mystery ("the light that shineth in darkness") which furnish the keys to the understanding of directly experienced realities. The fragmentariness and brevity of life, united to its dignity, is a mystery which is made meaningful by the promise of a fulfillment which is beyond the capacity of man. His sin and guilt are intolerable burdens without forgiveness which is available to those who acknowledge rather than hide the ultimate predicament of man. The dramas of man's history can be explained by discerning various patterns in history, but they are too multifarious to fit into any of these patterns. They can be fitted into a framework of meaning only if the meaning has a penumbra of mystery. The mystery consists of a power and a love beyond our comprehension which overrules these various historical dramas. The relation of the creative power to the redemptive love is also a mystery without the apprehension of which the whole of history falls into meaninglessness.

The hope that both the individual and the total drama of life will end in "the forgiveness of sins: the resurrection of the body: and the life everlasting," is thus the natural fruit of the faith that there is a height and depth of reality in God in which the individual in his uniqueness and freedom has a reality which the coherences of nature and reason do not assure him, and in which the endless dramatic variations of his collective life also have significance ultimately.

The Biblical faith makes the affirmation not only about the divine life which assures significance to selfhood and its dramas, but it also insists that the God who is powerful enough to bring the fragmentary dialogues and dramas to a conclusion also has a power of love which is able to overcome the recalcitrance of human sin.

There is no way of making either this faith or this hope "rational" by analyzing the coherences of nature and of reason. Efforts to do so inevitably result in some form of very attenuated philosophy and religion in which some of the experienced incoherences of life are obscured by the philosophically established coherences. One may only validate such a faith inferentially by calling attention to the fact that it answers the ultimate problems of the human self, which also has an incredible but directly experienced freedom, unity of freedom and finiteness and the capacity to elaborate endless dramas which do not fit into any pattern of nature and reason.

The fact that the individual feels his life to be fragmentary, despite its fulfillments in the community and in the historical process, proves that the self has a dimension which does not fit into any historical correlations. The divine power and love which gives this freedom meaning must be apprehended by faith. The reality of this power can be proved inferentially in the sense that any meaning which is imparted to life, purely in terms of the historical flux, can be proved to be a too simple meaning.

The fact that the self's freedom is involved in the contradiction of self-love is established by any careful observation of human history. But the self does not acknowledge itself to be implicated in this contradiction except in an encounter with the God who stands above all the historical judgements.

The self is involved in, and creates, various historical dramas which are known to have meaning. But the meaning which is ascribed to them is usually too simple if the frame of meaning consists merely of the correlations which may be charted by historical observation. Certainly the idea of progress which was the regulative religious idea of the past decades is an inadequate principle of meaning for the realities of an age which faces the dangers of atomic wars.

The unity of the self in its freedom and its natural and historical structure certainly invalidates all hopes of heaven which rest upon the idea that an immortal "soul" or mind may escape a mortal body. Thus the Biblical faith and hope, which gives meaning to human existence, may be proved inferentially to be true, or to be more in accord with experienced facts than alternative faith and hopes. But these inferences can not force the self to commit itself to them. The Biblical faith must remain a commitment of the self rather than a conclusion of its mind. Only such a commitment can do justice to the self's freedom and its discontinuity with the coherences and structures of the temporal world. The commitment is not possible without the prerequisite of repentance because the darkness about the meaning of its existence is due not so much to the finiteness of the self's mind as to pretensions of its heart. It is because men "became vain in their imaginations, and their foolish heart was darkened" (Romans 1:21) that they found such great difficulty in recognizing the true Author and end of their existence and the Divine Judge of the actions.

The dramas of history contain many facts and sequences which must be rationally correlated. But the frame of meaning in which these facts and sequences are discerned must be apprehended by faith because it touches the realm of mystery beyond rational comprehension. The ultimate question always remains whether the mystery is so absolute as to annul the meaning of the historical drama or whether there is a key of meaning in the mystery, a "light that shineth in darkness," which clarifies, rather than annuls, all the strange and variegated dramas of human history.

PART II

THE NON-THEISTIC ALTERNATIVE

❧ 5 ❧

BERTRAND RUSSELL

A Free Man's Worship

To Dr. Faustus in his study Mephistophelis told the history of the Creation, saying,

> The endless praises of the choirs of angels had begun to grow wearisome; for, after all, did he not deserve their praise? Had he not given them endless joy? Would it not be more amusing to obtain undeserved praise, to be worshiped by beings whom he tortured? He smiled inwardly, and resolved that the great drama should be performed.
>
> For countless ages the hot nebula whirled aimlessly through space. At length it began to take shape, the central mass threw off planets, the planets cooled, boiling seas and burning mountains heaved and tossed, from black masses of cloud hot sheets of rain deluged the barely solid crust. And now the first germ of life grew in the depths of the ocean and developed rapidly in the fructifying warmth into vast forest trees, huge ferns springing from the damp mold, sea monsters breeding, fighting, devouring, and passing away. And from the monsters, as the play unfolded itself, Man was born, with the power of thought, the knowledge of good and evil, and the cruel thirst for worship. And Man saw that all is passing in this mad, monstrous world, that all is struggling to snatch, at any cost, a few brief moments of life before Death's inexorable decree. And Man said, "There is a hidden purpose, could we but fathom it, and the purpose is good; for we must reverence something, and in the visible world there is nothing worthy of reverence." And Man stood aside from the struggle, resolving that God intended harmony to come out of chaos by human efforts. And when he followed the instincts which God had transmitted to him from his ancestry of beasts of prey, he called it Sin, and asked God to forgive him. But he doubted whether he could be justly forgiven, until he invented a divine Plan by which God's wrath was to have been appeased. And seeing the present was bad, he made it yet worse, that thereby the future might be better. And he gave God thanks for the strength that enabled him to forgo even the joys that were possible. And

God smiled; and when he saw that Man had become perfect in renunciation and worship, he sent another sun through the sky, which crashed into Man's sun; and all returned again to nebula.

"Yes," he murmured, "it was a good play; I will have it performed again."

Such, in outline, but even more purposeless, more void of meaning, is the world which science presents for our belief. Amid such a world, if anywhere, our ideals henceforward must find a home. That man is the product of causes which had no prevision of the end they were achieving; that his origin, his growth, his hopes and fears, his loves and his beliefs, are but the outcome of accidental collocations of atoms; that no fire, no heroism, no intensity of thought and feeling, can preserve an individual life beyond the grave; that all the labors of the ages, all the devotion, all the inspiration, all the noonday brightness of human genius, are destined to extinction in the vast death of the solar system, and that the whole temple of man's achievement must inevitably be buried beneath the debris of a universe in ruins—all these things, if not quite beyond dispute, are yet so nearly certain that no philosophy which rejects them can hope to stand. Only within the scaffolding of these truths, only in the firm foundation of unyielding despair, can the soul's habitation henceforth be safely built.

How, in such an alien and inhuman world, can so powerless a creature as man preserve his aspirations untarnished? A strange mystery it is that nature, omnipotent but blind, in the revolutions of her secular hurryings through the abysses of space, has brought forth at last a child, subject still to her power, but gifted with sight, with knowledge of good and evil, with the capacity of judging all the works of his unthinking mother. In spite of death, the mark and seal of the parental control, man is yet free, during his brief years, to examine, to criticize, to know, and in imagination to create. To him alone, in the world with which he is acquainted, this freedom belongs; and in this lies his superiority to the resistless forces that control his outward life.

The savage, like ourselves, feels the oppression of his impotence before the powers of nature; but having in himself nothing that he respects more than power, he is willing to prostrate himself before his gods, without inquiring whether they are worthy of his worship. Pathetic and very terrible is the long history of cruelty and torture, of degradation and human sacrifice, endured in the hope of placating the jealous gods: surely, the trembling believer thinks, when what is most precious has been freely given, their lust for blood must be appeased, and more will not be required. The religion of Moloch—as such creeds may be generically called—is in essence the cringing submission of the slave, who dare not, even in his heart, allow the thought that his master deserves no adulation. Since the independence of ideals is not yet acknowledged, power may be freely worshiped and receive an unlimited respect, despite its wanton infliction of pain.

But gradually, as morality grows bolder, the claim of the ideal world begins to be felt; and worship, if it is not to cease, must be given to gods of another kind than those created by the savage. Some, though they feel the demands of the ideal, will still consciously reject them, still urging that naked power is worthy of worship. Such is the attitude inculcated in God's answer to Job out of the whirlwind: the divine power and knowledge are paraded, but of the divine goodness there is no hint. Such also is the attitude of those who, in our own day, base their morality upon the struggle for survival, maintaining that the survivors are necessarily the fittest. But others, not content with an answer so repugnant to the moral sense, will adopt the position which we have become accustomed to regard as specially religious, maintaining that, in some hidden manner, the world of fact is really harmonious with the world of ideals. Thus man created God, all-powerful and all-good, the mystic unity of what is and what should be.

But the world of fact, after all, is not good; and, in submitting our judgment to it, there is an element of slavishness from which our thoughts must be purged. For in all things it is well to exalt the dignity of man, by freeing him as far as possible from the tyranny of nonhuman power. When we have realized that power is largely bad, that man, with his knowledge of good and evil, is but a helpless atom in a world which has no such knowledge, the choice is again presented to us: Shall we worship force, or shall we worship goodness? Shall our God exist and be evil, or shall he be recognized as the creation of our own conscience?

The answer to this question is very momentous and affects profoundly our whole morality. The worship of force, to which Carlyle and Nietzsche and the creed of militarism have accustomed us, is the result of failure to maintain our own ideals against a hostile universe: it is itself a prostrate submission to evil, a sacrifice of our best to Moloch. If strength indeed is to be respected, let us respect rather the strength of those who refuse that false "recognition of facts" which fails to recognize that facts are often bad. Let us admit that, in the world we know, there are many things that would be better otherwise, and that the ideals to which we do and must adhere are not realized in the realm of matter. Let us preserve our respect for truth, for beauty, for the ideal of perfection which life does not permit us to attain, though none of these things meet with the approval of the unconscious universe. If power is bad, as it seems to be, let us reject it from our hearts. In this lies man's true freedom: in determination to worship only the God created by our own love of the good, to respect only the heaven which inspires the insight of our best moments. In action, in desire, we must submit perpetually to the tyranny of outside forces; but in thought, in aspiration, we are free, free from our fellow men, free from the petty planet on which our bodies impotently crawl, free even, while we live, from the tyranny of death. Let us learn, then, that energy of faith which enables us to live constantly in the vision of the good; and let us descend, in action, into the world of fact, with that vision always before us.

When first the opposition of fact and ideal grows fully visible, a spirit of fiery revolt, of fierce hatred of the gods, seems necessary to the assertion of freedom. To defy with Promethean constancy a hostile universe, to keep its evil always in view, always actively hated, to refuse no pain that the malice of power can invent, appears to be the duty of all who will not bow before the inevitable. But indignation is still a bondage, for it compels our thoughts to be occupied with an evil world; and in the fierceness of desire from which rebellion springs there is a kind of self-assertion which it is necessary for the wise to overcome. Indignation is a submission of our thoughts but not of our desires; the Stoic freedom in which wisdom consists is found in the submission of our desires but not of our thoughts. From the submission of our desires springs the virtue of resignation; from the freedom of our thoughts springs the whole world of art and philosophy, and the vision of beauty by which, at last, we half reconquer the reluctant world. But the vision of beauty is possible only to unfettered contemplation, to thoughts not weighted by the load of eager wishes; and thus freedom comes only to those who no longer ask of life that it shall yield them any of those personal goods that are subject to the mutations of time.

Although the necessity of renunciation is evidence of the existence of evil, yet Christianity, in preaching it, has shown a wisdom exceeding that of the Promethean philosophy of rebellion. It must be admitted that, of the things we desire, some, though they prove impossible, are yet real goods; others, however, as ardently longed for, do not form part of a fully purified ideal. The belief that what must be renounced is bad, though sometimes false, is far less often false than untamed passion supposes; and the creed of religion, by providing a reason for proving that it is never false, has been the means of purifying our hopes by the discovery of many austere truths.

But there is in resignation a further good element: even real goods, when they are unattainable, ought not to be fretfully desired. To every man comes, sooner or later, the great renunciation. For the young, there is nothing unattainable; a good thing desired with the whole force of a passionate will, and yet impossible, is to them not credible. Yet, by death, by illness, by poverty, or by the voice of duty, we must learn, each one of us, that the world was not made for us, and that, however beautiful may be the things we crave, Fate may nevertheless forbid them. It is the part of courage, when misfortune comes, to bear without repining the ruin of our hopes, to turn away our thoughts from vain regrets. This degree of submission to power is not only just and right: it is the very gate of wisdom.

But passive renunciation is not the whole of wisdom; for not by renunciation alone can we build a temple for the worship of our own ideals. Haunting foreshadowings of the temple appear in the realm of imagination, in music, in architecture, in the untroubled kingdom of reason, and in the golden sunset magic of lyrics, where beauty shines and glows, remote from the touch of sorrow, remote from the fear of change, remote from the failures and disenchantments of the world of fact. In the contemplation of these things the vision of heaven will shape itself in our hearts, giving at once a touchstone to judge the world about us and an inspiration by which to fashion to our needs whatever is not incapable of serving as a stone in the sacred temple.

Except for those rare spirits that are born without sin, there is a cavern of darkness to be traversed before that temple can be entered. The gate of the cavern is despair, and its floor is paved with the gravestones of abandoned hopes. There self must die; there the eagerness, the greed of untamed desire, must be slain, for only so can the soul be freed from the empire of Fate. But out of the cavern, the Gate of Renunciation leads again to the daylight of wisdom, by whose radiance a new insight, a new joy, a new tenderness, shine forth to gladden the pilgrim's heart.

When, without the bitterness of impotent rebellion, we have learned both to resign ourselves to the outward rule of Fate and to recognize that the nonhuman world is unworthy of our worship, it becomes possible at last so to transform and refashion the unconscious universe, so to transmute it in the crucible of imagination, that a new image of shining gold replaces the old idol of clay. In all the multiform facts of the world—in the visual shapes of trees and mountains and clouds, in the events of the life of man, even in the very omnipotence of death—the insight of creative idealism can find the reflection of a beauty which its own thoughts first made. In this way mind asserts its subtle mastery over the thoughtless forces of nature. The more evil the material with which it deals, the more thwarting to untrained desire, the greater is its achievement in inducing the reluctant rock to yield up its hidden treasures, the prouder its victory in compelling the opposing forces to swell the pageant of its triumph. Of all the arts, tragedy is the proudest, the most triumphant; for it builds its

shining citadel in the very center of the enemy's country, on the very summit of his highest mountain; from its impregnable watchtowers, his camps
and arsenals, his columns and forts, are all revealed; within its walls the
free life continues, while the legions of death and pain and despair, and
all the servile captains of tyrant Fate, afford the burghers of that dauntless
city new spectacles of beauty. Happy those sacred ramparts, thrice happy
the dwellers on that all-seeing eminence. Honor to those brave warriors
who, through countless ages of warfare, have preserved for us the priceless heritage of liberty and have kept undefiled by sacrilegious invaders
the home of the unsubdued.

But the beauty of tragedy does but make visible a quality which, in
more or less obvious shapes, is present always and everywhere in life. In
the spectacle of death, in the endurance of intolerable pain, and in the
irrevocableness of a vanished past, there is a sacredness, an overpowering
awe, a feeling of the vastness, the depth, the inexhaustible mystery of
existence, in which, as by some strange marriage of pain, the sufferer is
bound to the world by bonds of sorrow. In these moments of insight, we
lose all eagerness of temporary desire, all struggling and striving for petty
ends, all care for the little trivial things that, to a superficial view, make
up the common life of day by day; we see, surrounding the narrow raft
illumined by the flickering light of human comradeship, the dark ocean
on whose rolling waves we toss for a brief hour; from the great night without, a chill blast breaks in upon our refuge; all the loneliness of humanity
amid hostile forces is concentrated upon the individual soul, which must
struggle alone, with what of courage it can command, against the whole
weight of a universe that cares nothing for its hopes and fears. Victory, in
this struggle with the powers of darkness, is the true baptism into the
glorious company of heroes, the true initiation into the overmastering
beauty of human existence. From that awful encounter of the soul with
the outer world, renunciation, wisdom, and charity are born; and with
their birth a new life begins. To take into the inmost shrine of the soul
the irresistible forces whose puppets we seem to be—death and change,
the irrevocableness of the past, and the powerlessness of man before the
blind hurry of the universe from vanity to vanity—to feel these things and
know them is to conquer them.

This is the reason why the past has such magical power. The beauty
of its motionless and silent pictures is like the enchanted purity of late
autumn, when the leaves, though one breath would make them fall, still
glow against the sky in golden glory. The past does not change or strive;
like Duncan, after life's fitful fever it sleeps well; what was eager and
grasping, what was petty and transitory, has faded away; the things that
were beautiful and eternal shine out of it like stars in the night. Its beauty,

to a soul not worthy of it, is unendurable; but to a soul which has conquered Fate it is the key of religion.

The life of man, viewed outwardly, is but a small thing in comparison with the forces of nature. The slave is doomed to worship Time and Fate and Death, because they are greater than anything he finds in himself, and because all his thoughts are of things which they devour. But, great as they are, to think of them greatly, to feel their passionless splendor, is greater still. And such thought makes us free men; we no longer bow before the inevitable in Oriental subjection, but we absorb it and make it a part of ourselves. To abandon the struggle for private happiness, to expel all eagerness of temporary desire, to burn with passion for eternal things—this is emancipation, and this is the free man's worship. And this liberation is effected by contemplation of Fate; for Fate itself is subdued by the mind which leaves nothing to be purged by the purifying fire of time.

United with his fellow men by the strongest of all ties, the tie of a common doom, the free man finds that a new vision is with him always, shedding over every daily task the light of love. The life of man is a long march through the night, surrounded by invisible foes, tortured by weariness and pain, toward a goal that few can hope to reach, and where none may tarry long. One by one, as they march, our comrades vanish from our sight, seized by the silent orders of omnipotent death. Very brief is the time in which we can help them, in which their happiness or misery is decided. Be it ours to shed sunshine on their path, to lighten their sorrows by the balm of sympathy, to give them the pure joy of a never-tiring affection, to strengthen failing courage, to instill faith in hours of despair. Let us not weigh in grudging scales their merits and demerits, but let us think only of their need—of the sorrows, the difficulties, perhaps the blindness, that make the misery of their lives; let us remember that they are fellow sufferers in the same darkness, actors in the same tragedy with ourselves. And so, when their day is over, when their good and their evil have become eternal by the immortality of the past, be it ours to feel that, where they suffered, where they failed, no deed of ours was the cause; but wherever a spark of the divine fire kindled in their hearts, we were ready with encouragement, with sympathy, with brave words in which high courage glowed.

Brief and powerless is man's life; on him and all his race the slow, sure doom falls pitiless and dark. Blind to good and evil, reckless of destruction, omnipotent matter rolls on its relentless way; for man, condemned today to lose his dearest, tomorrow himself to pass through the gate of darkness, it remains only to cherish, ere yet the blow fall, the lofty thoughts that ennoble his little day; disdaining the coward terrors of the slave of Fate, to worship at the shrine that his own hands have built; un-dismayed by the empire of chance, to preserve a mind free from the wan-ton tyranny that rules his outward life; proudly defiant of the irresistible forces that tolerate, for a moment, his knowledge and his condemnation, to sustain alone, a weary but unyielding Atlas, the world that his own ideals have fashioned despite the trampling march of unconscious power.

≥ 6 ≤

JULIAN HUXLEY

The Creed of
a Scientific Humanist

I believe that life can be worth living. I believe this in spite of pain, squalor, cruelty, unhappiness, and death. I do not believe that it is necessarily worth living, but only that for most people it can be.

I also believe that man, as individual, as group, and collectively as mankind, can achieve a satisfying purpose in existence. I believe this in spite of frustration, aimlessness, frivolity, boredom, sloth, and failure. Again I do not believe that a purpose inevitably inheres in the universe or in our existence, or that mankind is bound to achieve a satisfying purpose, but only that such a purpose can be found.

I believe that there exists a scale or hierarchy of values, ranging from simple physical comforts up to the highest satisfactions of love, esthetic enjoyment, intellect, creative achievement, virtue. I do not believe that these are absolute, or transcendental in the sense of being vouchsafed by some external power or divinity: they are the product of human nature interacting with the outer world. Nor do I suppose that we can grade every valuable experience into an accepted order, any more than I can say whether a beetle is a higher organism than a cuttlefish or a herring. But just as it can unhesitatingly be stated that there are general grades of biological organization, and that a beetle *is* a higher organism than a sponge, or a human being than a frog, so I can assert, with the general consensus of civilized human beings, that there is a higher value in Dante's *Divine Comedy* than in a popular hymn, in the scientific activity of Newton or Darwin than in solving a crossword puzzle, in the fullness of love than in sexual gratification, in selfless than in purely self-regarding activities—although each and all can have their value of a sort.

I do not believe that there is any absolute of truth, beauty, morality, or virtue, whether emanating from an external power or imposed by an internal standard. But this does not drive me to the curious conclusion, fashionable in certain quarters, that truth and beauty and goodness do not exist, or that there is no force or value in them.

I believe that there are a number of questions that it is no use our asking, because they can never be answered. Nothing but waste, worry, or unhappiness is caused by trying to solve insoluble problems. Yet some people seem determined to try. I recall the story of the philosopher and the theologian. The two were engaged in disputation and the theologian used the old quip about a philosopher resembling a blind man, in a dark room, looking for a black cat—which wasn't there. "That may be," said the philosopher: "but a theologian would have found it."

Even in matters of science, we must learn to ask the right questions. It seemed an obvious question to ask how animals inherit the result of their parents' experience, and enormous amounts of time and energy have been spent on trying to give an answer to it. It is, however, no good asking the question, for the simple reason that no such inheritance of acquired characters exists. The chemists of the eighteenth century, because they asked themselves the question, "What substance is involved in the process of burning?" became involved in the mazes of the phlogiston theory; they had to ask, "What sort of process is burning?" before they could see that it did not involve a special substance, but was merely a particular case of chemical combination.

When we come to what are usually referred to as fundamentals, the difficulty of not asking the wrong kind of question is much increased. Among most African tribes, if a person dies, the only question asked is, "Who caused his death, and by what form of magic?" The idea of death from natural causes is unknown. Indeed, the life of the less civilized half of mankind is largely based on trying to find an answer to a wrong question, "What magical forces or powers are responsible for good or bad fortune, and how can they be circumvented or propitiated?"

I do not believe in the existence of a god or gods. The conception of divinity seems to me, though built up out of a number of real elements of experience, to be a false one, based on the quite unjustifiable postulate that there must be some more or less personal power in control of the world. We are confronted with forces beyond our control, with incomprehensible disasters, with death; and also with ecstasy, with a mystical sense of union with something greater than our ordinary selves, with sudden conversion to a new way of life, with the burden of guilt and sin and of ways in which these burdens may be lifted. In theistic religions, all these elements of actual experience have been woven into a unified body of belief and practice, in relation to the fundamental postulate of the existence of a god or gods.

I believe this fundamental postulate to be nothing more than the result of asking a wrong question, "Who or what rules the universe?" So far as we can see, it rules itself, and indeed the whole analogy with a country and its ruler is false. Even if a god does exist behind or above the universe as we experience it, we can have no knowledge of such a power: the actual gods of historical religions are only the personifications of impersonal facts of nature and of facts of our inner mental life. Though we can answer the question, "What are the Gods of actual religions?" we can only do so by dissecting them into their components and showing their divinity to be a figment of human imagination, emotion, and rationalization. The question, "What is the nature of God?" we cannot answer, since we have no means of knowing whether such a being exists or not.

Similarly with immortality. With our present faculties, we have no means of giving a categorical answer to the question whether we survive death, much less the question of what any such life after death will be like. That being so, it is a waste of time and energy to devote ourselves to the problem of achieving salvation in the life to come. However, just as the idea of God is built out of bricks of real experience, so too is the idea of salvation. If we translate salvation into terms of this world, we find that it means achieving harmony between different parts of our nature, including its subconscious depths and its rarely touched heights, and also achieving some satisfactory relation of adjustment between ourselves and the outer world, including not only the world of nature, but the social world of man. I believe it to be possible to "achieve salvation" in this sense, and right to aim at doing so, just as I believe it possible and valuable to achieve a sense of union with something bigger than our ordinary selves, even if that something be not a god but an extension of our narrow core to include in a single grasp ranges of outer experience and inner nature on which we do not ordinarily draw.

But if God and immortality be repudiated, what is left? That is the question usually thrown at the atheist's head. The orthodox believer likes to think that nothing is left. That, however, is because he has only been accustomed to think in terms of his orthodoxy.

In point of fact, a great deal is left.

That is immediately obvious from the fact that many men and women have led active, or self-sacrificing, or noble, or devoted lives without any belief in God or immortality. Buddhism in its uncorrupted form has no such belief, nor did the great nineteenth-century agnostics, nor do the orthodox Russian Communists, nor did the Stoics. Of course, the unbelievers have often been guilty of selfish or wicked actions; but so have the believers. And in any case that is not the fundamental point. The point is that without these beliefs men and women may yet possess the mainspring of full and purposive living, and just as strong a sense that existence can be worth while as is possible to the most devout believers.

I would say that this is much more readily possible today than in any previous age. The reason lies in the advances of science.

No longer are we forced to accept the external catastrophes and miseries of existence as inevitable or mysterious; no longer are we obliged to live in a world without history, where change is only meaningless. Our ancestors saw an epidemic as an act of divine punishment: to us it is a challenge to be overcome, since we know its causes and that it could be controlled or prevented. The understanding of infectious disease is entirely due to scientific advance. So, to take a very recent happening, is our understanding of the basis of nutrition, which holds out new possibilities of health and energy to the human race. So is our understanding of earthquakes and storms: if we cannot control them, we at least do not have to fear them as evidence of God's anger.

Some, at least, of our internal miseries can be lightened in the same way. Through knowledge derived from psychology, children can be prevented from growing up with an abnormal sense of guilt, and so making life a burden both to themselves and to those with whom they come into contact. We are beginning to understand the psychological roots of irrational fear and irrational cruelty: some day we shall be able to make the world a brighter place by preventing their appearance.

The ancients had no history worth mentioning. Human existence in the present was regarded as a degradation from that of the original Golden Age. Down even to the nineteenth century, what was known of human history was regarded by the nations of the West as an essentially meaningless series of episodes sandwiched into the brief space between the Creation and the Fall, a few thousand years ago, and the Second Coming and the Last Judgment, which might be on us at any moment, and in any case could not be pushed back for more than a few thousand years into the future. In this perspective, a millennium was almost an eternity. With such an outlook, no wonder life seemed, to the great mass of humanity, "nasty, brutish and short," its miseries and shortcomings merely bewildering unless illuminated by the illusory light of religion.

Today, human history merges back into prehistory, and prehistory again into biological evolution. Our time-scale is profoundly altered. A thousand years is a short time for prehistory, which thinks in terms of hundreds of thousands of years, and an insignificant time for evolution, which deals in ten-million-year periods. The future is extended equally with the past: if it took over a thousand million years for primeval life to generate man, man and his descendents have at least an equal allowance of time before them.

Most important of all, the new history has a basis of hope. Biological evolution has been appallingly slow and appallingly wasteful. It has been cruel, it has generated the parasites and the pests as well as the more agreeable types. It has led life up innumerable blind alleys. But in spite of this, it has achieved progress. In a few lines, whose number has steadily diminished with time, it has avoided the cul-de-sac of mere specialization and arrived at a new level of organization, more harmonious and more efficient, from which it could again launch out toward greater control, greater knowledge, and greater independence. Progress is, if you will, all-round specialization. Finally, but one line was left which was able to achieve further progress: all the others had led up blind alleys. This was the line leading to the evolution of the human brain.

This at one bound altered the perspective of evolution. Experience could now be handed down from generation to generation; deliberate purpose could be substituted for the blind sifting of selection; change could be speeded up ten-thousandfold. In man evolution could become conscious. Admittedly it is far from conscious yet, but the possibility is there, and it has at least been consciously envisaged.

Seen in this perspective, human history represents but the tiniest portion of the time man has before him; it is only the first ignorant and clumsy gropings of the new type, born heir to so much biological history. Attempts at a general philosophy of history are seen in all their futility— as if someone whose acquaintance with man as a species were limited to a baby one year old should attempt a general account of the human mind and soul. The constant setbacks, the lack of improvement in certain respects for over two thousand years, are seen to be phenomena as natural as the tumbles of a child learning to walk, or the deflection of a sensitive boy's attention by the need of making a living.

The broad facts remain. Life had progressed, even before man was first evolved. Life progressed further by evolving man. Man has progressed during the half million or so years from the first Hominidae, even during the ten thousand years since the final amelioration of climate after the Ice Age. And the potentialities of progress which are revealed, once his eyes have been opened to the evolutionary vista, are unlimited.

At last we have an optimistic, instead of a pessimistic, theory of this world and our life upon it. Admittedly the optimism cannot be facile, and must be tempered with reflection on the length of time involved, on the hard work that will be necessary, on the inevitable residuum of accident and unhappiness that will remain. Perhaps we had better call it a melioris-

tic rather than an optimistic view: but at least it preaches hope and inspires to action.

I believe very definitely that it is among human personalities that
there exist the highest and most valuable achievements of the universe—
or at least the highest and most valuable achievements of which we have
or, apparently, can have knowledge. That means that I believe that the
State exists for the development of individual lives, not individuals for the
development of the State.

But I also believe that the individual is not an isolated, separate
thing. An individual is a transformer of matter and experience: it is a system of relations between its own basis and the universe, including other
individuals. An individual may believe that he should devote himself entirely to a cause, even sacrifice himself to it—his country, truth, art, love.
It is in the devotion or the sacrifice that he becomes most himself, it is because of the devotion or sacrifice of individuals that causes become of
value. But of course the individual must in many ways subordinate himself to the community—only not to the extent of believing that in the
community resides any virtue higher than that of the individuals which
compose it.

The community provides the machinery for the existence and development of individuals. There are those who deny the importance of social
machinery, who assert that the only important thing is a change of heart,
and that the right machinery is merely a natural consequence of the right
inner attitude. This appears to me mere solipsism. Different kinds of social machinery predispose to different inner attitudes. The most admirable
machinery is useless if the inner life is unchanged: but social machinery
can affect the fullness and quality of life. Social machinery can be de-
vised to make war more difficult, to promote health, to add interest to life.
Let us not despise machinery in our zeal for fullness of life, any more
than we should dream that machinery can ever automatically grind out
perfection of living.

I believe in diversity. Every biologist knows that human beings differ in their hereditary outfits, and therefore in the possibilities that they
can realize. Psychology is showing us how different are the types that
jostle each other on the world's streets. No amount of persuasion or education can make the extravert really understand the introvert, the verbalist
understand the lover of handicraft, the nonmathematical or nonmusical
person understand the passion of the mathematician or the musician. We
can try to forbid certain attitudes of mind. We could theoretically breed
out much of human variety. But this would be a sacrifice. Diversity is not
only the salt of life, but the basis of collective achievement. And the com-

plement of diversity is tolerance and understanding. This does not mean rating all values alike. We must protect society against criminals: we must struggle against what we think wrong. But just as in our handling of the criminal we should try to reform rather than merely to punish, so we must try to understand why we judge others' actions as wrong, which implies trying to understand the workings of our own minds, and discounting our own prejudices.

Finally, I believe that we can never reduce our principles to any few simple terms. Existence is always too various and too complicated. We must supplement principles with faith. And the only faith that is both concrete and comprehensive is in life, its abundance and its progress. My final belief is in life.

❧ 7 ❧

ALBERT CAMUS

The Absurdity
of Human Existence

ABSURDITY AND SUICIDE

There is but one truly serious philosophical problem, and that is suicide. Judging whether life is or is not worth living amounts to answering the fundamental question of philosophy. All the rest—whether or not the world has three dimensions, whether the mind has nine or twelve categories—comes afterwards. These are games; one must first answer. And if it is true, as Nietzsche claims, that a philosopher, to deserve our respect, must preach by example, you can appreciate the importance of that reply, for it will precede the definitive act. These are facts the heart can feel; yet they call for careful study before they become clear to the intellect.

If I ask myself how to judge that this question is more urgent than that, I reply that one judges by the actions it entails. I have never seen anyone die for the ontological argument. Galileo, who held a scientific truth of great importance, abjured it with the greatest of ease as soon as it endangered his life. In a certain sense, he did right.[1] That truth was not worth the stake. Whether the earth or the sun revolves around the other is a matter of profound indifference. To tell the truth, it is a futile question. On the other hand, I see many people die because they judge that life is not worth living. I see others paradoxically getting killed for the ideas or illusions that give them a reason for living (what is called a reason for living is also an excellent reason for dying). I therefore conclude that the meaning of life is the most urgent of questions. How to answer

---

1. From the point of view of the relative value of truth. On the other hand, from the point of view of virile behavior, this scholar's fragility may well make us smile.

it? On all essential problems (I mean thereby those that run the risk of leading to death or those that intensify the passion of living) there are probably but two methods of thought: the method of La Palisse and the method of Don Quixote. Solely the balance between evidence and lyricism can allow us to achieve simultaneously emotion and lucidity. In a subject at once so humble and so heavy with emotion, the learned and classical dialectic must yield, one can see, to a more modest attitude of mind deriving at one and the same time from common sense and understanding.

Suicide has never been dealt with except as a social phenomenon. On the contrary, we are concerned here, at the outset, with the relationship between individual thought and suicide. An act like this is prepared within the silence of the heart, as is a great work of art. The man himself is ignorant of it. One evening he pulls the trigger or jumps. Of an apartment-building manager who had killed himself I was told that he had lost his daughter five years before, that he had changed greatly since, and that that experience had "undermined" him. A more exact word cannot be imagined. Beginning to think is beginning to be undermined. Society has but little connection with such beginnings. The worm is in man's heart. That is where it must be sought. One must follow and understand this fatal game that leads from lucidity in the face of existence to flight from light. . . .

But if it is hard to fix the precise instant, the subtle step when the mind opted for death, it is easier to deduce from the act itself the consequences it implies. In a sense, and as in melodrama, killing yourself amounts to confessing. It is confessing that life is too much for you or that you do not understand it. Let's not go too far in such analogies, however, but rather return to everyday words. It is merely confessing that that "is not worth the trouble." Living, naturally, is never easy. You continue making the gestures commanded by existence for many reasons, the first of which is habit. Dying voluntarily implies that you have recognized, even instinctively, the ridiculous character of that habit, the absence of any profound reason for living, the insane character of that daily agitation, and the uselessness of suffering.

What, then, is that incalculable feeling that deprives the mind of the sleep necessary to life? A world that can be explained even with bad reasons is a familiar world. But, on the other hand, in a universe suddenly divested of illusions and lights, man feels an alien, a stranger. His exile is without remedy since he is deprived of the memory of a lost home or the hope of a promised land. This divorce between man and his life, the actor and his setting, is properly the feeling of absurdity. All healthy men having thought of their own suicide, it can be seen, without further explanation, that there is a direct connection between this feeling and the longing for death.

The subject of this essay is precisely this relationship between the absurd and suicide, the exact degree to which suicide is a solution to the absurd. The principle can be established that for a man who does not cheat, what he believes to be true must determine his action. Belief in the absurdity of existence must then dictate his conduct. It is legitimate to wonder, clearly and without false pathos, whether a conclusion of this importance requires forsaking as rapidly as possible an incomprehensible condition. I am speaking, of course, of men inclined to be in harmony with themselves.

Stated clearly, this problem may seem both simple and insoluble. But it is wrongly assumed that simple questions involve answers that are no less simple and that evidence implies evidence. *A priori* and reversing the terms of the problem, just as one does or does not kill oneself, it seems that there are but two philosophical solutions, either yes or no. This would be too easy. But allowance must be made for those who, without concluding, continue questioning. Here I am only slightly indulging in irony: this is the majority. I notice also that those who answer "no" act as if they thought "yes." As a matter of fact, if I accept the Nietzschean criterion, they think "yes" in one way or another. On the other hand, it often happens that those who commit suicide were assured of the meaning of life. These contradictions are constant. It may even be said that they have never been so keen as on this point where, on the contrary, logic seems so desirable. It is a commonplace to compare philosophical theories and the behavior of those who profess them. . . . Schopenhauer is often cited, as a fit subject for laughter, because he praised suicide while seated at a well-set table. This is no subject for joking. That way of not taking the tragic seriously is not so grievous, but it helps to judge a man.

In the face of such contradictions and obscurities must we conclude that there is no relationship between the opinion one has about life and the act one commits to leave it? Let us not exaggerate in this direction. In a man's attachment to life there is something stronger than all the ills in the world. The body's judgment is as good as the mind's, and the body shrinks from annihilation. We get into the habit of living before acquiring the habit of thinking. In that race which daily hastens us toward death, the body maintains its irreparable lead. In short, the essence of that contradiction lies in what I shall call the act of eluding because it is both less and more than diversion in the Pascalian sense. Eluding is the invariable game. The typical act of eluding, the fatal evasion that constitutes the third theme of this essay, is hope. Hope for another life one must "deserve" or trickery of those who live not for life itself but for some great idea that will transcend it, refine it, give it a meaning, and betray it. . . .

ABSURDITY AND MEANING

All great deeds and all great thoughts have a ridiculous beginning. Great works are often born on a street-corner or in a restaurant's revolving door. So it is with absurdity. The absurd world more than others derives its nobility from that abject birth. In certain situations, replying "nothing" when asked what one is thinking about may be pretense in a man. Those who are loved are well aware of this. But if that reply is sincere, if it symbolizes that odd state of soul in which the void becomes eloquent, in which the chain of daily gestures is broken, in which the heart vainly seeks the link that will connect it again, then it is as it were the first sign of absurdity.

It happens that the stage sets collapse. Rising, streetcar, four hours in the office or the factory, meal, streetcar, four hours of work, meal, sleep, and Monday Tuesday Wednesday Thursday Friday and Saturday according to the same rhythm—this path is easily followed most of the time. But one day the "why" arises and everything begins in that weariness tinged with amazement. "Begins"—this is important. Weariness comes at the end of the acts of a mechanical life, but at the same time it inaugurates the impulse of consciousness. It awakens consciousness and provokes what follows. What follows is the gradual return into the chain or it is the definitive awakening. At the end of the awakening comes, in time, the consequence: suicide or recovery. In itself weariness has something sickening about it. Here, I must conclude that it is good. For everything begins with consciousness and nothing is worth anything except through it. . . .

At the heart of all beauty lies something inhuman, and these hills, the softness of the sky, the outline of these trees at this very minute lose the illusory meaning with which we had clothed them, henceforth more remote than a lost paradise. The primitive hostility of the world rises up to face us across millennia. For a second we cease to understand it because for centuries we have understood it in solely the images and designs that we had attributed to it beforehand, because henceforth we lack the power to make use of that artifice. The world evades us because it becomes itself again. That stage scenery masked by habit becomes again what it is. It withdraws at a distance from us. Just as there are days when under the familiar face of a woman, we see as a stranger her we had loved months or years ago, perhaps we shall come even to desire what suddenly leaves us so alone. But the time has not yet come. Just one thing: that denseness and that strangeness of the world is the absurd.

Men, too, secrete the inhuman. At certain moments of lucidity, the mechanical aspect of their gestures, their meaningless pantomine makes silly everything that surrounds them. A man is talking on the telephone behind a glass partition; you cannot hear him, but you see his incomprehensible dumb show: you wonder why he is alive. This discomfort in the face of man's own inhumanity, this incalculable tumble before the image of what we are, this "nausea," as a writer of today calls it, is also the absurd. Likewise, the stranger who at certain seconds comes to meet us in a mirror, the familiar and yet alarming brother we encounter in our own photographs is also the absurd.

I come at last to death and to the attitude we have toward it. On this point everything has been said and it is only proper to avoid pathos. Yet one will never be sufficiently surprised that everyone lives as if no one "knew." This is because in reality there is no experience of death. Properly speaking, nothing has been experienced but what has been lived and made conscious. Here, it is barely possible to speak of the experience of others' deaths. It is a substitute, an illusion, and it never quite convinces us. That melancholy convention cannot be persuasive. The horror comes in reality from the mathematical aspect of the event. If time frightens us, this is because it works out the problem and the solution comes afterward. All the pretty speeches about the soul will have their contrary convincingly proved, at least for a time. From this inert body on which a slap makes no mark the soul has disappeared. This elementary and definitive aspect of the adventure constitutes the absurd feeling. Under the fatal lighting of that destiny, its uselessness becomes evident. No code of ethics and no effort are justifiable a priori in the face of the cruel mathematics that command our condition. . . .

Understanding the world for a man is reducing it to the human, stamping it with his seal. The cat's universe is not the universe of the anthill. The truism "All thought is anthropomorphic" has no other meaning. Likewise, the mind that aims to understand reality can consider itself satisfied only by reducing it to terms of thought. If man realized that the universe like him can love and suffer, he would be reconciled. If thought discovered in the shimmering mirrors of phenomena eternal relations capable of summing them up and summing themselves up in a single principle, then would be seen an intellectual joy of which the myth of the blessed would be but a ridiculous imitation. That nostalgia for unity, that appetite for the absolute illustrates the essential impulse of the human drama. But the fact of that nostalgia's existence does not imply that it is to be immediately satisfied. . . .

With the exception of professional rationalists, today people despair of true knowledge. If the only significant history of human thought were to be written, it would have to be the history of its successive regrets and its impotences.

Of whom and of what indeed can I say: "I know that!" This heart within me I can feel, and I judge that it exists. This world I can touch, and I likewise judge that it exists. There ends all my knowledge, and the rest is construction. For if I try to seize this self of which I feel sure, if I try to define and to summarize it, it is nothing but water slipping through my fingers. I can sketch one by one all the aspects it is able to assume, all those likewise that have been attributed to it, this upbringing, this origin, this ardor of these silences, this nobility or this vileness. But aspects cannot be added up. This very heart which is mine will forever remain indefinable to me. Between the certainty I have of my existence and the content I try to give to that assurance, the gap will never be filled. Forever I shall be a stranger to myself. . . .

Hence the intelligence, too, tells me in its way that this world is absurd. . . . In this unintelligible and limited universe, man's fate henceforth assumes its meaning. A horde of irrationals has sprung up and surrounds him until his ultimate end. In his recovered and now studied lucidity, the feeling of the absurd becomes clear and definite. I said that the world is absurd, but I was too hasty. This world in itself is not reasonable, that is all that can be said. But what is absurd is the confrontation of this irrational and the wild longing for clarity whose call echoes in the human heart. The absurd depends as much on man as on the world. . . .

I don't know whether this world has a meaning that transcends it. But I know that I do not know that meaning and that it is impossible for me just now to know it. What can a meaning outside my condition mean to me? I can understand only in human terms. What I touch, what resists me—that is what I understand. And these two certainties—my appetite for the absolute and for unity and the impossibility of reducing this world to a rational and reasonable principle—I also know that I cannot reconcile them. What other truth can I admit without lying, without bringing in a hope I lack and which means nothing within the limits of my condition?

If I were a tree among trees, a cat among animals, this life would have a meaning, or rather this problem would not rise, for I should belong to this world. I should *be* this world to which I am now opposed by my whole consciousness and my whole insistence upon familiarity. This ridiculous reason is what sets me in opposition to all creation. I cannot cross it out with a stroke of the pen. What I believe to be true I must therefore preserve. What seems to me so obvious, even against me, I must support. And what constitutes the basis of that conflict, of that break between the world and my mind, but the awareness of it? If therefore I want to preserve it, I can through a constant awareness, ever revived, ever alert. This is what, for the moment, I must remember. . . .

Let us insist again on the method: it is a matter of persisting. At a certain point on his path the absurd man is tempted. History is not lacking in either religions or prophets, even without gods. He is asked to leap. All he can reply is that he doesn't fully understand, that it is not obvious. Indeed, he does not want to do anything but what he fully understands. He is assured that this is the sin of pride, but he does not understand the notion of sin; that perhaps hell is in store, but he has not enough imagination to visualize that strange future; that he is losing immortal life, but that seems to him an idle consideration. An attempt is made to get him to admit his guilt. He feels innocent. To tell the truth, that is all he feels—his irreparable innocence. This is what allows him everything. Hence, what he demands of himself is to live *solely* with what he knows, to accommodate himself to what is, and to bring in nothing that is not certain. He is told that nothing is. But this at least is a certainty. And it is with this that he is concerned: he wants to find out if it is possible to live *without appeal*. . . .

Before encountering the absurd, the everyday man lives with aims, a concern for the future or for justification (with regard to whom or what is not the question). He weighs his chances, he counts on "someday," his retirement or the labor of his sons. He still thinks that something in his life can be directed. In truth, he acts as if he were free, even if all the facts make a point of contradicting that liberty. But after the absurd, everything is upset. That idea that "I am," my way of acting as if everything has a meaning (even if, on occasion, I said that nothing has)—all that is given the lie in vertiginous fashion by the absurdity of a possible death. Thinking of the future, establishing aims for oneself, having preferences—all this presupposes a belief in freedom, even if one occasionally ascertains that one doesn't feel it. But at that moment I am well aware that that higher liberty, that freedom *to be*, which alone can serve as basis for a truth, does not exist. Death is there as the only reality. . . .

But at the same time the absurd man realizes that hitherto he was bound to that postulate of freedom on the illusion of which he was living. In a certain sense, that hampered him. To the extent to which he imagined a purpose to his life, he adapted himself to the demands of a purpose to be achieved and became the slave of his liberty. Thus I could not act otherwise than as the father (or the engineer or the leader of a nation, or the post-office subclerk) that I am preparing to be. . . .

The absurd enlightens me on this point: there is no future. Henceforth, this is the reason for my inner freedom. . . .

But what does life mean in such a universe? Nothing else for the moment but indifference to the future and a desire to use up everything that is given. Belief in the meaning of life always implies a scale of values, a choice, our preferences. Belief in the absurd, according to our definitions, teaches the contrary. But this is worth examining.

Knowing whether or not one can live *without appeal* is all that interests me. I do not want to get out of my depth. This aspect of life being given me, can I adapt myself to it? Now, faced with this particular concern, belief in the absurd is tantamount to substituting the quantity of experiences for the quality. If I convince myself that this life has no other aspect than that of the absurd, if I feel that its whole equilibrium depends on that perpetual opposition between my conscious revolt and the darkness in which it struggles, if I admit that my freedom has no meaning except in relation to its limited fate, then I must say that what counts is not the best of living but the most living. . . .

On the one hand the absurd teaches that all experiences are unimportant, and on the other it urges toward the greatest quantity of experiences. How, then, can one fail to do as so many of those men I was speaking of earlier—choose the form of life that brings us the most possible of that human matter, thereby introducing a scale of values that on the other hand one claims to reject?

But again it is the absurd and its contradictory life that teaches us. For the mistake is thinking that that quantity of experiences depends on the circumstances of our life when it depends solely on us. Here we have to be over-simple. To two men living the same number of years, the world always provides the same sum of experiences. It is up to us to be conscious of them. Being aware of one's life, one's revolt, one's freedom, and to the maximum, is living, and to the maximum. Where lucidity dominates, the scale of values becomes useless. . . .

THE MYTH OF SISYPHUS

The gods had condemned Sisyphus to ceaselessly rolling a rock to the top of a mountain, whence the stone would fall back of its own weight. They had thought with some reason that there is no more dreadful punishment than futile and hopeless labor.

If one believes Homer, Sisyphus was the wisest and most prudent of mortals. According to another tradition, however, he was disposed to practice the profession of highwayman. I see no contradiction in this. Opinions differ as to the reasons why he became the futile laborer of the underworld. To begin with, he is accused of a certain levity in regard to the gods. He stole their secrets. Ægina, the daughter of Æsopus, was carried off by Jupiter. The father was shocked by that disappearance and complained to Sisyphus. He, who knew of the abduction, offered to tell about it on condition that Æsopus would give water to the citadel of Corinth. To the celestial thunderbolts he preferred the benediction of water. He was punished for this in the underworld. Homer tells us also that Sisyphus had put Death in chains. Pluto could not endure the sight of his deserted, silent empire. He dispatched the god of war, who liberated Death from the hands of her conqueror.

It is said also that Sisyphus, being near to death, rashly wanted to test his wife's love. He ordered her to cast his unburied body into the middle of the public square. Sisyphus woke up in the underworld. And there, annoyed by an obedience so contrary to human love, he obtained from Pluto permission to return to earth in order to chastise his wife. But when he had seen again the face of this world, enjoyed water and sun, warm stones and the sea, he no longer wanted to go back to the infernal darkness. Recalls, signs of anger, warnings were of no avail. Many years more he lived facing the curve of the gulf, the sparkling sea, and the smiles of earth. A decree of the gods was necessary. Mercury came and seized the impudent man by the collar and, snatching him from his joys, led him forcibly back to the underworld, where his rock was ready for him.

You have already grasped that Sisyphus is the absurd hero. He *is*, as much through his passions as through his torture. His scorn of the gods, his hatred of death, and his passion for life won him that unspeakable penalty in which the whole being is exerted toward accomplishing nothing. This is the price that must be paid for the passions of this earth. Nothing is told us about Sisyphus in the underworld. Myths are made for the imagination to breathe life into them. As for this myth, one sees merely the whole effort of a body straining to raise the huge stone, to roll it and push it up a slope a hundred times over; one sees the face screwed up, the cheek tight against the stone, the shoulder bracing the clay-covered mass, the foot wedging it, the fresh start with arms outstretched, the wholly human security of two earth-clotted hands. At the very end of his long effort measured by skyless space and time without depth, the purpose is achieved. Then Sisyphus watches the stone rush down in a few moments toward that lower world whence he will have to push it up again toward the summit. He goes back down to the plain.

It is during that return, that pause, that Sisyphus interests me. A face that toils so close to stones is already stone itself! I see that man going back down with a heavy yet measured step toward the torment of which he will never know the end. That hour like a breathing-space which returns as surely as his suffering, that is the hour of consciousness.

At each of those moments when he leaves the heights and gradually sinks toward the lairs of the gods, he is superior to his fate. He is stronger than his rock.

If this myth is tragic, that is because its hero is conscious. Where would his torture be, indeed, if at every step the hope of succeeding upheld him? The workman of today works every day in his life at the same tasks, and this fate is no less absurd. But it is tragic only at the rare moments when it becomes conscious. Sisyphus, proletarian of the gods, powerless and rebellious, knows the whole extent of his wretched condition: it is what he thinks of during his descent. The lucidity that was to constitute his torture at the same time crowns his victory. There is no fate that cannot be surmounted by scorn.

If the descent is thus sometimes performed in sorrow, it can also take place in joy. This word is not too much. Again I fancy Sisyphus returning toward his rock, and the sorrow was in the beginning. When the images of earth cling too tightly to memory, when the call of happiness becomes too insistent, it happens that melancholy rises in man's heart: this is the rock's victory, this is the rock itself. The boundless grief is too heavy to bear. These are our nights of Gethsemane. But crushing truths perish from being acknowledged. Thus, Œdipus at the outset obeys fate without knowing it. But from the moment he knows, his tragedy begins. Yet at the same moment, blind and desperate, he realizes that the only bond linking him to the world is the cool hand of a girl. Then a tremendous remark rings out: "Despite so many ordeals, my advanced age and the nobility of my soul make me conclude that all is well." Sophocles' Œdipus, like Dostoevsky's Kirilov, thus gives the recipe for the absurd victory. Ancient wisdom confirms modern heroism.

One does not discover the absurd without being tempted to write a manual of happiness. "What! by such narrow ways—?" There is but one world, however. Happiness and the absurd are two sons of the same earth. They are inseparable. It would be a mistake to say that happiness necessarily springs from the absurd discovery. It happens as well that the feeling of the absurd springs from happiness. "I conclude that all is well," says Œdipus, and that remark is sacred. It echoes in the wild and limited universe of man. It teaches that all is not, has not been, exhausted. It drives out of this world a god who had come into it with dissatisfaction and a preference for futile sufferings. It makes of fate a human matter, which must be settled among men.

All Sisyphus' silent joy is contained therein. His fate belongs to him. His rock is his thing. Likewise, the absurd man, when he contemplates his torment, silences all the idols. In the universe suddenly restored to its

silence, the myriad wondering little voices of the earth rise up. Unconscious, secret calls, invitations from all the faces, they are the necessary
reverse and price of victory. There is no sun without shadow, and it is essential to know the night. The absurd man says yes and his effort will
henceforth be unceasing. If there is a personal fate, there is no higher destiny, or at least there is but one which he concludes is inevitable and de-
spicable. For the rest, he knows himself to be the master of his days. At
that subtle moment when man glances backward over his life, Sisyphus
returning toward his rock, in that slight pivoting he contemplates that series of unrelated actions which becomes his fate, created by him, combined under his memory's eye and soon sealed by his death. Thus, convinced of the wholly human origin of all that is human, a blind man eager
to see who knows that the night has no end, he is still on the go. The
rock is still rolling.

I leave Sisyphus at the foot of the mountain! One always finds one's
burden again. But Sisyphus teaches the higher fidelity that negates the
gods and raises rocks. He too concludes that all is well. This universe
henceforth without a master seems to him neither sterile nor futile. Each
atom of that stone, each mineral flake of that night-filled mountain, in itself forms a world. The struggle itself toward the heights is enough to fill
a man's heart. One must imagine Sisyphus happy.

8

KURT BAIER

The Meaning of Life

Tolstoy, in his autobiographical work, "A Confession," reports how, when he was fifty and at the height of his literary success, he came to be obsessed by the fear that life was meaningless.

> At first I experienced moments of perplexity and arrest of life, as though I did not know what to do or how to live; and I felt lost and became dejected. But this passed, and I went on living as before. Then these moments of perplexity began to recur oftener and oftener, and always in the same form. They were always expressed by the questions: What is it for? What does it lead to? At first it seemed to me that these were aimless and irrelevant questions. I thought that it was all well known, and that if I should ever wish to deal with the solution it would not cost me much effort; just at present I had no time for it, but when I wanted to, I should be able to find the answer. The questions however began to repeat themselves frequently, and to demand replies more and more insistently; and like drops of ink always falling on one place they ran together into one black blot.[1]

A Christian living in the Middle Ages would not have felt any serious doubts about Tolstoy's questions. To him it would have seemed quite certain that life had a meaning and quite clear what it was. The medieval Christian world picture assigned to man a highly significant, indeed the central part in the grand scheme of things. The universe was made for the express purpose of providing a stage on which to enact a drama starring Man in the title role.

To be exact, the world was created by God in the year 4004 B.C. Man was the last and the crown of this creation, made in the likeness of

---

1. Count Leo Tolstoy, "A Confession," reprinted in *A Confession, The Gospel in Brief, and What I Believe*, No. 229, The World's Classics (London: Geoffrey Cumberlege, 1940).

God, placed in the Garden of Eden on earth, the fixed centre of the universe, round which revolved the nine heavens of the sun, the moon, the planets and the fixed stars, producing as they revolved in their orbits the heavenly harmony of the spheres. And this gigantic universe was created for the enjoyment of man, who was originally put in control of it. Pain and death were unknown in paradise. But this state of bliss was not to last. Adam and Eve ate of the forbidden tree of knowledge, and life on this earth turned into a death-march through a vale of tears. Then, with the birth of Jesus, new hope came into the world. After He had died on the cross, it became at least possible to wash away with the purifying water of baptism some of the effects of Original Sin and to achieve salvation. That is to say, on condition of obedience to the law of God, man could now enter heaven and regain the state of everlasting, deathless bliss, from which he had been excluded because of the sin of Adam and Eve.

To the medieval Christian the meaning of human life was therefore perfectly clear. The stretch on earth is only a short interlude, a temporary incarceration of the soul in the prison of the body, a brief trial and test, fated to end in death, the release from pain and suffering. What really matters, is the life after the death of the body. One's existence acquires meaning not by gaining what this life can offer but by saving one's immortal soul from death and eternal torture, by gaining eternal life and everlasting bliss.

The scientific world picture which has found ever more general acceptance from the beginning of the modern era onwards is in profound conflict with all this. At first, the Christian conception of the world was discovered to be erroneous in various important details. The Copernican theory showed up the earth as merely one of several planets revolving round the sun, and the sun itself was later seen to be merely one of many fixed stars each of which is itself the nucleus of a solar system similar to our own. Man, instead of occupying the centre of creation, proved to be merely the inhabitant of a celestial body no different from millions of others. Furthermore, geological investigations revealed that the universe was not created a few thousand years ago, but was probably millions of years old.

Disagreements over details of the world picture, however, are only superficial aspects of a much deeper conflict. The appropriateness of the whole Christian outlook is at issue. For Christianity, the world must be regarded as the "creation" of a kind of Superman, a person possessing all the human excellences to an infinite degree and none of the human weaknesses, Who has made man in His image, a feeble, mortal, foolish copy of Himself. In creating the universe, God acts as a sort of playwright-cum-legislator-cum-judge-cum-executioner. In the capacity of playwright,

He creates the historical world process, including man. He erects the stage and writes, in outline, the plot. He creates the *dramatis personae* and watches over them with the eye partly of a father, partly of the law. While on stage, the actors are free to extemporise, but if they infringe the divine commandments, they are later dealt with by their creator in His capacity of judge and executioner.

Within such a framework, the Christian attitudes towards the world are natural and sound: it is natural and sound to think that all is arranged for the best even if appearances belie it; to resign oneself cheerfully to one's lot; to be filled with awe and veneration in regard to anything and everything that happens; to want to fall on one's knees and worship and praise the Lord. These are wholly fitting attitudes within the framework of the world view just outlined. And this world view must have seemed wholly sound and acceptable because it offered the best explanation which was then available of all the observed phenomena of nature.

As the natural sciences developed, however, more and more things in the universe came to be explained without the assumption of a supernatural creator. Science, moreover, could explain them better, that is, more accurately and more reliably. The Christian hypothesis of a supernatural maker, whatever other needs it was capable of satisfying, was at any rate no longer indispensable for the purpose of explaining the existence or occurrence of anything. In fact, scientific explanations do not seem to leave any room for this hypothesis. The scientific approach demands that we look for a natural explanation of anything and everything. The scientific way of looking at and explaining things has yielded an immensely greater measure of understanding of, and control over, the universe than any other way. And when one looks at the world in this scientific way, there seems to be no room for a personal relationship between human beings and a supernatural perfect being ruling and guiding men. Hence many scientists and educated men have come to feel that the Christian attitudes towards the world and human existence are inappropriate. They have become convinced that the universe and human existence in it are without a purpose and therefore devoid of meaning.[2]

1. THE EXPLANATION OF THE UNIVERSE

Such beliefs are disheartening and unplausible. It is natural to keep looking for the error that must have crept into our arguments. And if an error

2. See e.g. Edwyn Bevan, *Christianity*, pp. 211–227. See also H. J. Paton, *The Modern Predicament* (London: George Allen and Unwin Ltd., 1955) pp. 103–116, 374.

has crept in, then it is most likely to have crept in with science. For before the rise of science, people did not entertain such melancholy beliefs, while the scientific world picture seems literally to force them on us.

There is one argument which seems to offer the desired way out. It runs somewhat as follows. Science and religion are not really in conflict. They are, on the contrary, mutually complementary, each doing an entirely different job. Science gives provisional, if precise, explanations of small parts of the universe, religion gives final and over-all, if comparatively vague, explanations of the universe as a whole. The objectionable conclusion, that human existence is devoid of meaning, follows only if we use scientific explanations where they do not apply, namely, where total explanations of the whole universe are concerned.⸰

After all, the argument continues, the scientific world picture is the inevitable outcome of rigid adherence to scientific method and explanation, but scientific, that is, causal explanations from their very nature are incapable of producing real illumination. They can at best tell us *how* things are or have come about, but never *why*. They are incapable of making the universe intelligible, comprehensible, meaningful to us. They represent the universe as meaningless, not because it *is* meaningless, but because scientific explanations are not designed to yield answers to investigations into the why and wherefore, into the meaning, purpose, or point of things. Scientific explanations (this argument continues) began, harmlessly enough, as partial and provisional explanations of the movement of material bodies, in particular the planets, within the general framework of the medieval world picture. Newton thought of the universe as a clock made, originally wound up, and occasionally set right by God. His laws of motion only revealed the ways in which the heavenly machinery worked. Explaining the movement of the planets by these laws was analogous to explaining the machinery of a watch. Such explanations showed *how* the thing worked, but not *what it was for* or *why* it existed. Just as the explanation of how a watch works can help our understanding of the watch only if, in addition, we assume that there is a watchmaker who has designed it for a purpose, made it, and wound it up, so the Newtonian explanation of the solar system helps our understanding of it only on the similar assumption that there is some divine artificer who has designed and made this heavenly clockwork for some purpose, has wound it up, and perhaps even occasionally sets it right, when it is out of order.

Socrates, in the Phaedo complained that only explanations of a thing showing the good or purpose for which it existed could offer a *real* expla-

---

3. See for instance, L. E. Elliott-Binns, *The Development of English Theology in the Later Nineteenth Century* (London: Longmans, Green & Co., 1952) pp. 30–33.

nation of it. He rejected the kind of explanation we now call "causal" as no more than mentioning "that without which a cause could not be a cause," that is, as merely a necessary condition, but not the *real* cause, the real explanation.[4] In other words, Socrates held that *all* things can be explained in two different ways: either by mentioning merely a necessary condition, or by giving the *real* cause. The former is not an elucidation of the explicandum, not really a help in understanding it, in grasping its "why" and "wherefore."

This Socratic view, however, is wrong. It is not the case that there are two kinds of explanation for everything, one partial, preliminary, and not really clarifying, the other full, final, and illuminating. The truth is that these two kinds of explanation are equally explanatory, equally illuminating, and equally full and final, but that they are appropriate for different kinds of explicanda.

When in an uninhabited forest we find what looks like houses, paved streets, temples, cooking utensils, and the like, it is no great risk to say that these things are the ruins of a deserted city, that is to say, of something man-made. In such a case, the appropriate explanation is teleological, that is, in terms of the purposes of the builders of that city. On the other hand, when a comet approaches the earth, it is similarly a safe bet that, unlike the city in the forest, it was not manufactured by intelligent creatures and that, therefore, a teleological explanation would be out of place, whereas a causal one is suitable.

It is easy to see that in some cases causal, and in others teleological explanations are appropriate. A small satellite circling the earth may or may not have been made by man. We may never know which is the true explanation, but either hypothesis is equally explanatory. It would be wrong to say that only a teleological explanation can *really* explain it. Either explanation would yield complete clarity although, of course, only one can be true. Teleological explanation is only one of several that are possible.

It may indeed be strictly correct to say that the question "*Why* is there a satellite circling the earth?" can only be answered by a teleological explanation. It may be true that "Why?"—questions can really be used properly only in order to elicit *someone's reasons for* doing something. If this is so, it would explain our dissatisfaction with causal answers to "Why?"—questions. But even if it is so, it does not show that "Why is the satellite there?" *must be answered by a teleological explanation.* It shows only that either it must be so answered or it must not be asked.

4. See "Phaedo" (*Five Dialogues* by Plato, Everyman's Library No. 456) para. 99, p. 189.

The question "Why have you stopped beating your wife?" can be answered only by a teleological explanation, but if you have never beaten her, it is an improper question. Similarly, if the satellite is not man-made, "Why is there a satellite?" is improper since it implies an origin it did not have. Natural science can indeed only tell us *how* things in nature have come about and not *why*, but this is so not because something else can tell us the *why* and *wherefore,* but because there is none.

There is, however, another point which has not yet been answered. The objection just stated was that causal explanations did not even set out to answer the crucial question. We ask the question "Why?" but science returns an answer to the question "How?" It might now be conceded that this is no ground for a complaint, but perhaps it will instead be said that causal explanations do not give complete or full answers even to that latter question. In causal explanations, it will be objected, the existence of one thing is explained by reference to its cause, but this involves asking for the cause of that cause, and so on, ad infinitum. There is no resting place which is not as much in need of explanation as what has already been explained. Nothing at all is ever fully and completely explained by this sort of explanation.

Leibniz has made this point very persuasively. "Let us suppose a book of the elements of geometry to have been eternal, one copy always having been taken down from an earlier one; it is evident that, even though a reason can be given for the present book out of a past one, nevertheless, out of any number of books, taken in order, going backwards, we shall never come upon *a full* reason; though we might well always wonder why there should have been such books from all time—why there were books at all, and why they were written in this manner. What is true of books is true also of the different states of the world; for what follows is in some way copied from what precedes . . . And so, however far you go back to earlier states, you will never find in those states *a full reason* why there should be any world rather than none, and why it should be such as it is."[5]

However, a moment's reflection will show that if any type of explanation is merely preliminary and provisional, it is teleological explanation, since it presupposes a background which itself stands in need of explanation. If I account for the existence of the man-made satellite by saying that it was made by some scientists for a certain purpose, then such an explanation can clarify the existence of the satellite only if I assume that there existed materials out of which the satellite was made, and scientists

---

5. "On the Ultimate Origination of Things" (*The Philosophical Writings of Leibniz,* Everyman's Library No. 905) p. 32.

who made it for some purpose. It therefore does not matter what type of explanation we give, whether causal or teleological: either type, any type of explanation, will imply the existence of something by reference to which the explicandum can be explained. And this in turn must be accounted for in the same way, and so on for ever.

But is not God a necessary being? Do we not escape the infinite regress as soon as we reach God? It is often maintained that, unlike ordinary intelligent beings, God is eternal and necessary; hence His existence, unlike theirs, is not in need of explanation. For what is it that creates the vicious regress just mentioned? It is that, if we accept the principle of sufficient reason (that there must be an explanation for the existence of anything and everything the existence of which is not logically necessary, but merely contingent[6]), the existence of all the things referred to in any explanation requires itself to be explained. If, however, God is a logically necessary being, then His existence requires no explanation. Hence the vicious regress comes to an end with God.

Now, it need not be denied that God is a necessary being in some sense of that expression. In one of these senses, I, for instance, am a necessary being: it is impossible that I should not exist, because it is self-refuting to say "I do not exist." The same is true of the English language and of the universe. It is self-refuting to say "There is no such thing as the English language" because this sentence is in the English language, or "There is no such thing as the universe" because whatever there is, *is* the universe. It is impossible that these things should not in fact exist since it is impossible that we should be mistaken in thinking that they exist. For what possible occurrence could even throw doubt on our being right on these matters, let alone show that we are wrong? I, the English language, and the universe, are necessary beings, simply in the sense in which all is necessarily true which has been *proved* to be true. The occurrence of utterances such as "I exist," "The English language exists" and "The universe exists" is in itself sufficient proof of their truth. These remarks are therefore necessarily true, hence the things asserted to exist are necessary things.

But this sort of necessity will not satisfy the principle of sufficient reason, because it is only hypothetical or consequential necessity.[7] *Given that* someone says "I exist," then it is logically impossible that *he* should not exist. Given the evidence we have, the English language and the

6. See "Monadology" (*The Philosophical Writings of Leibniz,* Everyman's Library No. 905) para. 32–38, pp. 8–10.
7. To borrow the useful term coined by Professor D. A. T. Gasking of Melbourne University.

universe most certainly do exist. But there is no necessity about the evidence. On the principle of sufficient reason, we must explain the existence of the evidence, for its existence is not logically necessary.

In other words, the only sense of "necessary being" capable of terminating the vicious regress is "logically necessary being," but it is no longer seriously in dispute that the notion of a logically necessary being is self-contradictory.[8] Whatever can be conceived of as existing can equally be conceived of as not existing.

However, even if per impossible, there were such a thing as a logically necessary being, we could still not make out a case for the superiority of teleological over causal explanation. The existence of the universe cannot be explained in accordance with the familiar model of manufacture by a craftsman. For that model presupposes the existence of materials out of which the product is fashioned. God, on the other hand, must create the materials as well. Moreover, although we have a simple model of "creation out of nothing," for composers create tunes out of nothing, yet this is a great difference between creating *something to be sung,* and making the sounds which are a singing of it, or producing the piano on which to play it. Let us, however, waive all these objections and admit, for argument's sake, that creation out of nothing is conceivable. Surely, even so, no one can claim that it is the kind of explanation which yields the clearest and fullest understanding. Surely, to round off scientific explanations of the origin of the universe with creation out of nothing, does not add anything to our *understanding.* There may be merit of some sort in this way of speaking, but whatever it is, it is not greater clarity or explanatory power.[9]

What then, does all this amount to? Merely to the claim that scientific explanations are no worse than any other. All that has been shown is that all explanations suffer from the same defect: all involve a vicious infinite regress. In other words, no type of human explanation can help us to unravel the ultimate, unanswerable mystery. Christian ways of looking at things may not be able to render the world any more lucid than science can, but at least they do not pretend that there are no impenetrable mysteries. On the contrary, they point out untiringly that the claims of science to be able to elucidate everything are hollow. They remind us that science

8. See e.g. J. J. C. Smart, "The Existence of God," reprinted in *New Essays in Philosophical Theology,* ed. by A. Flew and A. MacIntyre (London: S.C.M. Press, 1957) pp. 35–39.

9. That creation out of nothing is not a clarificatory notion becomes obvious when we learn that "in the philosophical sense" it does not imply creation at a particular time. The universe could be regarded as a creation out of nothing even if it had no beginning. See e.g. E. Gilson, *The Christian Philosophy of St. Thomas Aquinas* (London: Victor Gollancz Ltd. 1957) pp. 147–155 and E. L. Mascall, *Via Media* (London: Longmans, Green & Co., 1956) pp. 28 ff.

is not merely limited to the exploration of a tiny corner of the universe but that, however far out probing instruments may eventually reach, we can never even approach the answers to the last questions: "Why is there a world at all rather than nothing?" and "Why is the world such as it is and not different?" Here our finite human intellect bumps against its own boundary walls.

Is it true that scientific explanations involve an infinite vicious regress? Are scientific explanations really only provisional and incomplete? The crucial point will be this. Do *all* contingent truths call for explanation? Is the principle of sufficient reason sound? Can scientific explanations never come to a definite end? It will be seen that with a clear grasp of the nature and purpose of explanation we can answer these questions.[10]

Explaining something to someone is making him understand it. This involves bringing together in his mind two things, a model which is accepted as already simple and clear, and that which is to be explained, the explicandum, which is not so. Understanding the explicandum is seeing that it belongs to a range of things which could legitimately have been expected by anyone familiar with the model and with certain facts.

There are, however, two fundamentally different positions which a person may occupy relative to some explicandum. He may not be familiar with any model capable of leading him to expect the phenomenon to be explained. Most of us, for instance, are in that position in relation to the phenomena occurring in a good seance. With regard to other things people will differ. Someone who can play chess, already understands chess, already has such a model. Someone who has never seen a game of chess has not. He sees the moves on the board but he cannot understand, cannot follow, cannot make sense of what is happening. Explaining the game to him is giving him an explanation, is making him understand. He can understand or follow chess moves only if he can see them as conforming to a model of a chess game. In order to acquire such a model, he will, of course, need to know the constitutive rules of chess, that is, the permissible moves. But that is not all. He must know that a normal game of chess is a competition (not all games are) between two people, each trying to win, and he must know what it is to win at chess: to manoeuvre the opponent's king into a position of check-mate. Finally, he must acquire some knowledge of what is and what is not conducive to winning: the tactical rules or canons of the game.

10. In what follows I have drawn heavily on the work of Ryle and Toulmin. See for instance G. Ryle, *The Concept of Mind* (London: Hutchinson's University Library, 1949) pp. 56–60 etc. and his article, "If, So, and Because," in *Philosophical Analysis* by Max Black, and S. E. Toulmin, *Introduction to the Philosophy of Science* (London: Hutchinson's University Library, 1953).

A person who has been given such an explanation and who has mastered it—which may take quite a long time—has now reached understanding, in the sense of the ability to follow each move. A person cannot in that sense understand merely one single move of chess and no other. If he does not understand any other moves, we must say that he has not yet mastered the explanation, that he does not really understand the single move either. If he has mastered the explanation, then he understands all those moves which he can see as being in accordance with the model of the game inculcated in him during the explanation.

However, even though a person who has mastered such an explanation will understand many, perhaps most, moves of any game of chess he cares to watch, he will not necessarily understand them all, as some moves of a player may not be in accordance with his model of the game. White, let us say, at his fifteenth move, exposes his queen to capture by Black's knight. Though in accordance with the constitutive rules of the game, this move is nevertheless perplexing and calls for explanation, because it is not conducive to the achievement by White of what must be assumed to be his aim: to win the game. The queen is a much more valuable piece than the knight against which he is offering to exchange.

An onlooker who has mastered chess may fail to understand this move, be perplexed by it, and wish for an explanation. Of course he may fail to be perplexed, for if he is a very inexperienced player he may not *see* the disadvantageousness of the move. But there is such a need whether anyone sees it or not. The move *calls for* explanation because to anyone who knows the game it must appear to be incompatible with the model which we have learnt during the explanation of the game, and by reference to which we all explain and understand normal games.

However, the required explanation of White's 15th move is of a very different kind. What is needed now is not the acquisition of an explanatory model, but the removal of the real or apparent incompatibility between the player's move and the model of explanation he has already acquired. In such a case the perplexity can be removed only on the assumption that the incompatibility between the model and the game is merely apparent. As our model includes a presumed aim of both players, there are the following three possibilities: (a) White has made a mistake: he has overlooked the threat to his queen. In that case, the explanation is that White thought his move conducive to his end, but it was not. (b) Black has made a mistake: White set a trap for him. In that case, the explanation is that Black thought White's move was not conducive to White's end, but it was. (c) White is not pursuing the end which any chess player may be presumed to pursue: he is not trying to win his game. In that case, the explanation is that White has made a move which he knows is not conducive to the end of winning his game because, let us say, he wishes to please Black who is his boss.

Let us now set out the differences and similarities between the two types of understanding involved in these two kinds of explanation. I shall call the first kind "model"—understanding and explaining, respectively, because both involve the use of a model by reference to which understanding and explaining is effected. The second kind I shall call "unvexing," because the need for this type of explanation and understanding arises only when there is a perplexity arising out of the incompatibility of the model and the facts to be explained.

The first point is that unvexing presupposes model-understanding, but not vice versa. A person can neither have nor fail to have unvexing-understanding of White's fifteenth move at chess, if he does not already have model-understanding of chess. Obviously, if I don't know how to play chess, I shall fail to have model-understanding of White's fifteenth move. But I can neither fail to have nor, of course, can I have unvexing-understanding of it, for I cannot be perplexed by it. I merely fail to have model-understanding of this move as, indeed, of any other move of chess. On the other hand, I may well have model-understanding of chess without having unvexing-understanding of every move. That is to say, I may well know how to play chess without understanding White's fifteenth move. A person cannot fail to have unvexing-understanding of the move unless he is vexed or perplexed by it, hence he cannot even fail to have unvexing-understanding unless he already has model-understanding. It is not true that one either understands or fails to understand. On certain occasions, one neither understands nor fails to understand.

The second point is that there are certain things which cannot call for unvexing-explanations. No one can for instance call for an unvexing-explanation of White's first move, which is Pawn to King's Four. For no one can be perplexed or vexed by this move. Either a person knows how to play chess or he does not. If he does, then he must understand this move, for if he does not understand it, he has not yet mastered the game. And if he does not know how to play chess, then he cannot yet have, or fail to have, unvexing-understanding, he cannot therefore need an unvexing-explanation. Intellectual problems do not arise out of ignorance, but out of insufficient knowledge. An ignoramus is puzzled by very little. Once a student can see problems, he is already well into the subject.

The third point is that model-understanding implies being able, without further thought, to have model-understanding of a good many other things, unvexing-understanding does not. A person who knows chess and therefore has model-understanding of it, must understand a good many chess moves, in fact all except those that call for unvexing-explanations.

If he claims that he can understand White's first move, but no others, then he is either lying or deceiving himself or he really does not understand any move. On the other hand, a person who, after an unvexing-explanation, understands White's fifteenth move, need not be able, without further explanation, to understand Black's or any other further move which calls for unvexing-explanation.

What is true of explaining deliberate and highly stylized human behaviour such as playing a game of chess is also true of explaining natural phenomena. For what is characteristic of natural phenomena, that they recur in essentially the same way, that they are, so to speak, repeatable, is also true of chess games, as it is not of games of tennis or cricket. There is only one important difference: man himself has invented and laid down the rules of chess, as he has not invented or laid down the "rules or laws governing the behaviour of things." This difference between chess and phenomena is important, for it adds another way to the three already mentioned,[11] in which a perplexity can be removed by an unvexing-explanation, namely, by abandoning the original explanatory model. This is, of course, not possible in the case of games of chess, because the model for chess is not a "construction" on the basis of the already existing phenomena of chess, but an invention. The person who first thought up the model of chess could not have been mistaken. The person who first thought of a model explaining some phenomenon could have been mistaken.

Consider an example. We may think that the following phenomena belong together: the horizon seems to recede however far we walk towards it; we seem to be able to see further the higher the mountain we climb; the sun and moon seem every day to fall into the sea on one side but to come back from behind the mountains on the other side without being any the worse for it. We may explain these phenomena by two alternative models: (a) that the earth is a large disc; (b) that it is a large sphere. However, to a believer in the first theory there arises the following perplexity: how is it that when we travel long enough towards the horizon in any one direction, we do eventually come back to our starting point without ever coming to the edge of the earth? We may at first attempt to "save" the model by saying that there is only an apparent contradiction. We may say either that the model does not require us to come to an edge, for it may be possible only to walk round and round on the flat surface. Or we may say that the person must have walked over the edge without noticing it, or perhaps that the travellers are all lying. Alternatively, the fact that our model is "constructed" and not invented or laid down en-

11. See above, p. 10, points (a)–(c).

ables us to say, what we could not do in the case of chess, that the model is inadequate or unsuitable. We can choose another model which fits all the facts, for instance, that the earth is round. Of course, then we have to give an unvexing-explanation for why it *looks* flat, but we are able to do that.

We can now return to our original question, "Are scientific explanations true and full explanations or do they involve an infinite regress, leaving them for ever incomplete?"

Our distinction between model- and unvexing-explanations will help here. It is obvious that only those things which are perplexing *call for* and *can be given* unvexing-explanations. We have already seen that in disposing of one perplexity, we do not necessarily raise another. On the contrary, unvexing-explanations truly and completely explain what they set out to explain, namely, how something is possible which, on our explanatory model, seemed to be impossible. There can therefore be no infinite regress here. Unvexing-explanations are real and complete explanations.

Can there be an infinite regress, then, in the case of model-explanations? Take the following example. European children are puzzled by the fact that their antipodean counterparts do not drop into empty space. This perplexity can be removed by substituting for their explanatory model another one. The European children imagine that throughout space there is an all-pervasive force operating in the same direction as the force that pulls them to the ground. We must, in our revised model, substitute for this force another acting everywhere in the direction of the centre of the earth. Having thus removed their perplexity by giving them an adequate model, we can, however, go on to ask *why* there should be such a force as the force of gravity, why bodies should "naturally," in the absence of forces acting on them, behave in the way stated in Newton's laws. And we might be able to give such an explanation. We might for instance construct a model of space which would exhibit as derivable from it what in Newton's theory are "brute facts." Here we would have a case of the brute facts of one theory being explained within the framework of another, more general theory. And it is a sound methodological principle that we should continue to look for more and more general theories.

Note two points, however. The first is that we must distinguish, as we have seen, between *the possibility* and *the necessity* of giving an explanation. Particular occurrences can be explained by being exhibited as instances of regularities, and regularities can be explained by being exhibited as instances of more general regularities. Such explanations make things clearer. They organize the material before us. They introduce order where previously there was disorder. But absence of this sort of explanation (model-explanation) does not leave us with a puzzle or perplexity, an intellectual restlessness or cramp. The unexplained things are not unintelligible, incomprehensible, or irrational. Some things, on the other hand, call for, require, demand an explanation. As long as we are without such an explanation, we are perplexed, puzzled, intellectually perturbed. We need an unvexing-explanation.

Now, it must be admitted that we may be able to construct a more general theory, from which, let us say, Newton's theory can be derived. This would further clarify the phenomena of motion and would be intellectually satisfying. But failure to do so would not leave us with an intellectual cramp. The facts stated in Newton's theory do not require, or stand in need of, unvexing-explanations. They could do so only if we already had another theory or model with which Newton's theory was incompatible. They could not do so, by themselves, prior to the establishment of such another model.

The second point is that there is an objective limit to which such explanations tend, and beyond which they are pointless. There is a very good reason for wishing to explain a less general by a more general theory. Usually, such a unification goes hand in hand with greater precision in measuring the phenomena which both theories explain. Moreover, the more general theory, because of its greater generality, can explain a wider range of phenomena including not only phenomena already explained by some other theories but also newly discovered phenomena, which the less general theory cannot explain. Now, the ideal limit to which such expansions of theories tend is an all-embracing theory which unifies all theories and explains all phenomena. Of course, such a limit can never be reached, since new phenomena are constantly discovered. Nevertheless, theories may be tending towards it. It will be remembered that the contention made against scientific theories was that there is no such limit because they involve an infinite regress. On that view, which I reject, there is no conceivable point at which scientific theories could be said to have explained the whole universe. On the view I am defending, there is such a limit, and it is the limit towards which scientific theories are actually tending. I claim that the nearer we come to this limit, the closer we are to a full and complete explanation of everything. For if we were to reach the limit, then though we could, of course, be left with a model which is itself unexplained and could be yet further explained by derivation from another model, there would be no need for, and no point in, such a further explanation. There would be no need for it, because any clearly defined model permitting us to expect the phenomena it is designed to explain offers full and complete explanations of these phenomena, however narrow the range. And while, at lower levels of generality, there is a good reason for providing more general models, since they further simplify, systematize, and organize the phenomena, this, which is the only reason for building more general theories, no longer applies once we reach the ideal limit of an all-embracing explanation.

It might be said that there is another reason for using different models: that they might enable us to discover new phenomena. Theories are not only instruments of explanation, but also of discovery. With this I agree, but it is irrelevant to my point: that *the needs of explanation* do not require us to go on for ever deriving one explanatory model from another.

It must be admitted, then, that in the case of model-explanations there is a regress, but it is neither vicious nor infinite. It is not vicious because, in order to explain a group of explicanda, a model-explanation *need* not itself be derived from another more general one. It gives a perfectly full and consistent explanation by itself. And the regress is not infinite, for there is a natural limit, an all-embracing model, which can explain all phenomena, beyond which it would be pointless to derive model-explanations from yet others.

What about our most serious question, "Why is there anything at all?" Sometimes, when we think about how one thing has developed out of another and that one out of a third, and so on back throughout all time, we are driven to ask the same question about the universe as a whole. We want to add up all things and refer to them by the name, "the world," and we want to know why the world exists and why there is not nothing instead. In such moments, the world seems to us a kind of bubble floating on an ocean of nothingness. Why should such flotsam be adrift in empty space? Surely, its emergence from the hyaline billows of nothingness is more mysterious even than Aphrodite's emergence from the sea. Wittgenstein expressed in these words the mystification we all feel: "Not *how* the world is, is the mystical, but *that* it is. The contemplation of the world *sub specie aeterni* is the contemplation of it as a limited whole. The feeling of the world as a limited whole is the mystical feeling."[12]

Professor J. J. C. Smart expresses his own mystification in these moving words:

"That anything should exist at all does seem to me a matter for the deepest awe. But whether other people feel this sort of awe, and whether they or I ought to is another question. I think we ought to. If so, the question arises: If 'Why should anything exist at all?' cannot be interpreted after the manner of the cosmological argument, that is, as an ab-

---

12. L. Wittgenstein, *Tractatus Logico-Philosophicus* (London: Routledge & Kegan Paul Ltd., 1922), Sect. 6.44–6.45.

surd request for the non-sensical postulation of a logically necessary being, what sort of question is it? What sort of question is this question 'Why should anything exist at all?' All I can say is that I do not yet know."[13]

It is undeniable that the magnitude and perhaps the very existence of the universe is awe-inspiring. It is probably true that it gives many people "the mystical feeling." It is also undeniable that our awe, our mystical feeling, aroused by contemplating the vastness of the world, is justified, in the same sense in which our fear is justified when we realize we are in danger. There is no more appropriate object for our awe or for the mystical feeling than the magnitude and perhaps the existence of the universe, just as there is no more appropriate object for our fear than a situation of personal peril. However, it does not follow from this that it is a good thing to cultivate, or indulge in, awe or mystical feelings, any more than it is necessarily a good thing to cultivate, or indulge in, fear in the presence of danger.

In any case, whether or not we ought to have or are justified in having a mystical feeling or a feeling of awe when contemplating the universe, having such a feeling is not the same as asking a meaningful question, although having it may well *incline us* to utter certain forms of words. Our question "Why is there anything at all?" may be no more than the expression of our feeling of awe or mystification, and not a meaningful question at all. Just as the feeling of fear may naturally but illegitimately give rise to the question "What sin have I committed?" so the feeling of awe or mystification may naturally but illegitimately lead to the question "Why is there anything at all?" What we have to discover, then, is whether this question makes sense or is meaningless.

Yes, of course, it will be said, it makes perfectly good sense. There is an undeniable fact and it calls for explanation. The fact is that the universe exists. In the light of our experience, there can be no possible doubt that something or other exists, and the claim that the universe exists commits us to no more than that. And surely this calls for explanation, because the universe must have originated somehow. Everything has an origin and the universe is no exception. Since the universe is the totality of things, it must have originated out of nothing. If it had originated out of something, even something as small as one single hydrogen atom, what has so originated could not be the whole universe, but only the universe minus the atom. And then the atom itself would call for explanation, for it too must have had an origin, and it must be *an origin out of nothing*. And

13. Op. cit. p. 46. See also Rudolf Otto, *The Idea of the Holy* (London: Geoffrey Cumberlege, 1952) esp. pp. 9–29.

how can anything originate out of nothing? Surely that calls for explanation.

However, let us be quite clear what is to be explained. There are two facts here, not one. The first is that the universe exists, which is undeniable. The second is that the universe must have originated out of nothing, and that is not undeniable. It is true that, *if it has originated at all,* then it must have originated out of nothing, or else it is not the universe that has originated. But need it have originated? Could it not have existed for ever?[14] It might be argued that nothing exists for ever, that everything has originated out of something else. That may well be true, but it is perfectly compatible with the fact that the universe is everlasting. We may well be able to trace the origin of any thing to the time when, by some transformation, it has developed out of some other thing, and yet it may be the case that no thing has its origin in nothing, and the universe has existed for ever. For even if every *thing* has a beginning and an end, the total of mass and energy may well remain constant.

Moreover, the hypothesis that the universe originated out of nothing is, empirically speaking, completely empty. Suppose, for argument's sake, that the annihilation of an object without remainder is conceivable. It would still not be possible for any hypothetical observer to ascertain whether space was empty or not. Let us suppose that *within the range of observation of our observer* one object after another is annihilated without remainder and that only one is left. Our observer could not then tell whether in remote parts of the universe, beyond his range of observation,

14. Contemporary theologians would admit that it cannot be proved that the universe must have had a beginning. They would admit that we know it only through revelation. (See footnote No. 9.) I take it more or less for granted that Kant's attempted proof of the Thesis in his First Antinomy of Reason [Immanuel Kant's *Critique of Pure Reason,* trans. by Norman Kemp Smith (London: Macmillan and Co. Ltd., 1950) pp. 396–402] is invalid. It rests on a premise which is false: that the completion of the infinite series of succession of states, which must have preceded the present state if the world has had no beginning, is logically impossible. We can persuade ourselves to think that this infinite series is logically impossible if we insist that it is a series which must, literally, be *completed.* For the verb "to complete," as normally used, implies an activity which, in turn, implies an agent who must have *begun* the activity at some time. If an infinite series is a whole that must be *completed* then, indeed, the world must have had a beginning. But that is precisely the question at issue. If we say, as Kant does at first, "that an eternity has elapsed," we do not feel the same impossibility. It is only when we take seriously the words "synthesis" and "completion," both of which suggest or imply "work" or "activity" and therefore "beginning," that it seems necessary that an infinity of successive states cannot have elapsed. [See also R. Crawshay-Williams, *Methods and Criteria of Reasoning* (London: Routledge & Kegan Paul, 1957) App. iv.]

objects are coming into being or passing out of existence. What, moreover, are we to say of the observer himself? Is he to count for nothing? Must we not postulate him away as well, if the universe is to have arisen out of nothing?

Let us, however, ignore all these difficulties and assume that the universe really has originated out of nothing. Even that does not prove that the universe has not existed for ever. If the universe can conceivably develop out of nothing, then it can conceivably vanish without remainder. And it can arise out of nothing again and subside into nothingness once more, and so on ad infinitum. Of course, "again" and "once more" are not quite the right words. The concept of time hardly applies to such universes. It does not make sense to ask whether one of them is earlier or later than, or perhaps simultaneous with, the other because we cannot ask whether they occupy the same or different spaces. Being separated from one another by "nothing," they are not separated from one another by "anything." We cannot therefore make any statements about their mutual spatio-temporal relations. It is impossible to distinguish between one long continuous universe and two universes separated by nothing. How, for instance, can we tell whether the universe including ourselves is not frequently annihilated and "again" reconstituted just as it was?

Let us now waive these difficulties as well. Let us suppose for a moment that we understand what is meant by saying that the universe originated out of nothing and that this has happened only once. Let us accept this as a fact. Does this fact call for explanation?

It does not call for an unvexing-explanation. That would be called for only if there were a perplexity due to the incompatibility of an accepted model with some fact. In our case, the fact to be explained is the origination of the universe out of nothing, hence there could not be such a perplexity, for we need not employ a model incompatible with this. If we had a model incompatible with our "fact," then that would be the wrong model and we would simply have to substitute another for it. The model we employ to explain the origin of the universe out of nothing could not be based on the similar origins of other things for, of course, there is nothing else with a similar origin.

All the same, it seems very surprising that something should have come out of nothing. It is contrary to the principle that every thing has an origin, that is, has developed out of something else. It must be admitted that there is this incompatibility. However, it does not arise because a well-established model does not square with an undeniable fact; it arises because a well-established model does not square with *an assumption* of which it is hard even to make sense and for which there is no evidence whatsoever. In fact, the only reason we have for making this assumption, is a simple logical howler: that because every thing has an origin, the universe must have an origin, too, except that, being the universe, it must have originated out of nothing. This is a howler, because it conceives of the universe as a big thing, whereas in fact it is the totality of things, that is, not a thing. That every thing has an origin does not entail that the totality of things has an origin. On the contrary, it strongly suggests that it has not. For to say that every thing has an origin implies that any given thing must have developed out of something else which in turn, being a thing, must have developed out of something else, and so forth. If we assume that every thing has an origin, we need not, indeed it is hard to see how we can, assume that the totality of things has an origin as well. There is therefore no perplexity, because we need not and should not assume that the universe has originated out of nothing.

If, however, in spite of all that has been said just now, someone still wishes to assume, contrary to all reason, that the universe has originated out of nothing, there would still be no perplexity, for then he would simply have to give up the principle which is incompatible with this assumption, namely, that no thing can originate out of nothing. After all, this principle *could* allow for exceptions. We have no proof that it does not. Again, there is no perplexity, because no incompatibility between our assumption and an inescapable principle.

But, it might be asked, do we not need a model-explanation of our supposed fact? The answer is No. We do not need such an explanation, for there could not possibly be a model for this origin other than this origin itself. We cannot say that origination out of nothing is like birth, or emergence, or evolution, or anything else we know for it is not like anything we know. In all these cases, there is *something* out of which the new thing has originated.

To sum up. The question, "Why is there anything at all?" looks like a perfectly sensible question modelled on "Why does *this* exist?" or "How has *this* originated?" It looks like a question about the origin of a thing. However, it is not such a question, for the universe is not a thing, but the totality of things. There is therefore no reason to assume that the universe has an origin. The very assumption that it has is fraught with contradictions and absurdities. If, nevertheless, it were true that the universe has originated out of nothing, then this would not call either for an unvexing- or a model-explanation. It would not call for the latter, because there could be no model of it taken from another part of our experience, since there is nothing analogous in our experience to origination out of nothing. It would not call for the former, because there can be no perplexity due to the incompatibility of a well-established model and an undeniable fact, since there is no undeniable fact and no well-established model. If, on the other hand, as is more probable, the universe has not originated at all, but is eternal, then the question why or how it has originated simply does not arise. There can then be no question about why anything at all exists, for it could not mean how or why the universe had originated, since ex hypothesi it has no origin. And what else could it mean?

Lastly, we must bear in mind that the hypothesis that the universe was made by God out of nothing only brings us back to the question who made God or how God originated. And if we do not find it repugnant to say that God is eternal, we cannot find it repugnant to say that the universe is eternal. The only difference is that we know for certain that the universe exists, while we have the greatest difficulty in even making sense of the claim that God exists.

To sum up. According to the argument examined, we must reject the scientific world picture because it is the outcome of scientific types of explanation which do not really and fully explain the world around us, but only tell us *how* things have come about, not *why,* and can give no answer to the ultimate question, why there is anything at all rather than nothing. Against this, I have argued that scientific explanations are real and full, just like the explanations of everyday life and of the traditional religions. They differ from those latter only in that they are more precise and more easily disprovable by the observation of facts.

My main points dealt with the question why scientific explanations were thought to be merely provisional and partial. The first main reason is the misunderstanding of the difference between teleological and causal explanations. It is first, and rightly, maintained that teleological explanations are answers to "Why?"-questions, while causal explanations are answers to "How?"-questions. It is further, and wrongly, maintained that, in order to obtain real and full explanations of anything, one must answer both "Why?" and "How?" questions. In other words, it is thought that all matters can and must be explained by both teleological and causal types of explanation. Causal explanations, it is believed, are merely provisional and partial, waiting to be completed by teleological explanations. Until a teleological explanation has been given, so the story goes, we have not *really* understood the explicandum. However, I have shown that both types are equally real and full explanations. The difference between them is merely that they are appropriate to different types of explicanda.

It should, moreover, be borne in mind that teleological explanations are not, in any sense, unscientific. They are rightly rejected in the natural sciences, not however because they are unscientific, but because no intelligences or purposes are found to be involved there. On the other hand, teleological explanations are very much in place in psychology, for we find intelligence and purpose involved in a good deal of human behaviour. It is not only not unscientific to give teleological explanations of deliberate human behaviour, but it would be quite unscientific to exclude them.

The second reason why scientific explanations are thought to be merely provisional and partial, is that they are believed to involve a vicious infinite regress. Two misconceptions have led to this important error. The first is the general misunderstanding of the nature of explanation, and in particular the failure to distinguish between the two types which I have called model- and unvexing-explanations, respectively. If one does not draw this distinction, it is natural to conclude that scientific explanations lead to a vicious infinite regress. For while it is true of those perplexing matters which are elucidated by unvexing-explanations that they are incomprehensible and cry out for explanation, it is not true that after an unvexing-explanation has been given, this itself is again capable, let alone in need of, a yet further explanation of the same kind. Conversely, while it is true that model-explanations of regularities can themselves be further explained by more general model-explanations, it is not true that, in the absence of such more general explanations, the less general are incomplete, hang in the air, so to speak, leaving the explicandum incomprehensible and crying out for explanation. The distinction between the two types of explanation shows us that an explicandum is either perplexing and incomprehensible, in which case an explanation of it *is necessary* for clarification and, when given, *complete*, or it is a regularity capable of being subsumed under a model, in which case a further explanation *is possible* and often profitable, but *not necessary* for clarification.

The second misconception responsible for the belief in a vicious infinite regress is the misrepresentation of scientific explanation *as essentially causal*. It has generally been held that, in a scientific explanation, the explicandum is the effect of some event, the cause, temporally prior to the explicandum. Combined with the principle of sufficient reason (the principle that anything is in need of explanation which might conceivably have been different from what it is), this error generates the nightmare of determinism. Since any event might have been different from what it was, acceptance of this principle has the consequence that *every* event must have a reason or explanation. But if the reason is itself an event *prior in time*, then every reason must have a reason preceding it, and so the infinite regress of explanation is necessarily tied to the time scale stretching infinitely back into the endless past. It is, however, obvious from our account that science is not primarily concerned with the forging of such causal chains. The primary object of the natural sciences is not historical at all. Natural science claims to reveal, not the beginnings of things, but their underlying reality. It does not dig up the past, it digs down into the structure of things existing here and now. Some scientists do allow themselves to speculate, and rather precariously at that, about origins. But their hard work is done on the structure of what exists now. In particular those explanations which are themselves further explained are not explanations linking event to event in a gapless chain reaching back to creation day, but generalisations of theories tending towards a unified theory.

2. THE PURPOSE OF MAN'S EXISTENCE

Our conclusion in the previous section has been that science is in principle able to give complete and real explanations of every occurrence and thing in the universe. This has two important corollaries: (i) Acceptance of the scientific world picture cannot be *one's reason for* the belief that the universe is unintelligible and therefore meaningless, though coming to accept it, after having been taught the Christian world picture, may well have been, in the case of many individuals, *the only or the main cause* of their belief that the universe and human existence are meaningless. (ii) It is not in accordance with reason to reject this pessimistic belief on the grounds that scientific explanations are only provisional and incomplete and must be supplemented by religious ones.

In fact, it might be argued that the more clearly we understand the explanations given by science, the more we are driven to the conclusion that human life has no purpose and therefore no meaning. The science of astronomy teaches us that our earth was not specially created about 6,000 years ago, but evolved out of hot nebulae which previously had whirled aimlessly through space for countless ages. As they cooled, the sun and the planets formed. On one of these planets at a certain time the circumstances were propitious and life developed. But conditions will not remain favourable to life. When our solar system grows old, the sun will cool, our planet will be covered with ice, and all living creatures will eventually perish. Another theory has it that the sun will explode and that the heat generated will be so great that all organic life on earth will be destroyed. That is the comparatively short history and prospect of life on earth. Altogether it amounts to very little when compared with the endless history of the inanimate universe.

Biology teaches us that the species man was not specially created but is merely, in a long chain of evolutionary changes of forms of life, the last link, made in the likeness not of God but of nothing so much as an ape. The rest of the universe, whether animate or inanimate, instead of serving the ends of man, is at best indifferent, at worst savagely hostile. Evolution to whose operation the emergence of man is due is a ceaseless battle among members of different species, one species being gobbled up by another, only the fittest surviving. Far from being the gentlest and most highly moral, man is simply the creature best fitted to survive, the most efficient if not the most rapacious and insatiable killer. And in this unplanned, fortuitous, monstrous, savage world man is madly trying to snatch a few brief moments of joy, in the short intervals during which he is free from pain, sickness, persecution, war or famine until, finally, his life is snuffed out in death. Science has helped us to know and understand this world, but what purpose or meaning can it find in it?

Complaints such as these do not mean quite the same to everybody, but one thing, I think, they mean to most people: science shows life to be meaningless, because life is without purpose. The medieval world picture provided life with a purpose, hence medieval Christians could believe that life had a meaning. The scientific account of the world takes away life's purpose and with it its meaning.

There are, however, two quite different senses of "purpose." Which one is meant? Has science deprived human life of purpose in both senses? And if not, is it a harmless sense, in which human existence has been robbed of purpose? Could human existence still have meaning if it did not have a purpose in that sense?

What are the two senses? In the first and basic sense, purpose is normally attributed only to persons or their behaviour as in "Did you have a purpose in leaving the ignition on?" In the second sense, purpose is normally attributed only to things, as in "What is the purpose of that gadget you installed in the workshop?" The two uses are intimately connected. We cannot attribute a purpose to a thing without implying that someone did something, in the doing of which he had some purpose, namely, to bring about the thing with the purpose. Of course, *his* purpose is not identical with *its* purpose. In hiring labourers and engineers and buying materials and a site for a factory and the like, the entrepreneur's purpose, let us say, is to manufacture cars, but the purpose of cars is to serve as a means of transportation.

There are many things that a man may do, such as buying and selling, hiring labourers, ploughing, felling trees, and the like, which it is foolish, pointless, silly, perhaps crazy, to do if one has no purpose in doing them. A man who does these things without a purpose is engaging in inane, futile pursuits. Lives crammed full with such activities devoid of purpose are pointless, futile, worthless. Such lives may indeed be dismissed as meaningless. But it should also be perfectly clear that acceptance of the scientific world picture does not force us to regard our lives as being without a purpose in this sense. Science has not only not robbed us of any purpose which we had before, but it has furnished us with enormously greater power to achieve these purposes. Instead of praying for rain or a good harvest or offspring, we now use ice pellets, artificial manure, or artificial insemination.

By contrast, having or not having a purpose, in the other sense, is value neutral. We do not think more or less highly of a thing for having or not having a purpose. "Having a purpose," in this sense, confers no kudos, "being purposeless" carries no stigma. A row of trees growing near a farm may or may not have a purpose: it may or may not be a windbreak, may or may not have been planted or deliberately left standing there in order to prevent the wind from sweeping across the fields. We do not in any way disparage the trees if we say they have no purpose, but have just grown that way. They are as beautiful, made of as good wood, as valuable, as if they had a purpose. And, of course, they break the wind just as well. The same is true of living creatures. We do not disparage a dog when we say that it has no purpose, is not a sheep dog or a watch dog or a rabbiting dog, but just a dog that hangs around the house and is fed by us.

Man is in a different category, however. To attribute to a human being a purpose in that sense is not neutral, let alone complimentary: it is offensive. It is degrading for a man to be regarded as merely serving a purpose. If, at a garden party, I ask a man in livery, "What is your purpose?" I am insulting him. I might as well have asked, "What are you for?" Such questions reduce him to the level of a gadget, a domestic animal, or perhaps a slave. I imply that we allot to him the tasks, the goals, the aims which he is to pursue; that his wishes and desires and aspirations and purposes are to count for little or nothing. We are treating him, in Kant's phrase, merely as a means to our ends, not as an end in himself.

The Christian and the scientific world pictures do indeed differ fundamentally on this point. The latter robs man of a purpose in this sense. It sees him as a being with no purpose allotted to him by anyone but himself. It robs him of any goal, purpose, or destiny appointed for him by any outside agency. The Christian world picture, on the other hand, sees man as a creature, a divine artefact, something halfway between a robot (manufactured) and an animal (alive), a homunculus, or perhaps Frankenstein, made in God's laboratory, with a purpose or task assigned him by his Maker.

However, lack of purpose in this sense does not in any way detract from the meaningfulness of life. I suspect that many who reject the scientific outlook because it involves the loss of purpose of life, and therefore meaning, are guilty of a confusion between the two senses of "purpose" just distinguished. They confusedly think that if the scientific world picture is true, then their lives must be futile because that picture implies that man has no purpose given him from without. But this is muddled thinking, for, as has already been shown, pointlessness, is implied only by purposelessness in the other sense, which is not at all implied by the scientific picture of the world. These people mistakenly conclude that there can be no purpose *in* life because there is no purpose *of* life; that *men* cannot themselves adopt and achieve purposes because *man*, unlike a robot or a watch dog, is not a creature with a purpose.[15]

However, not all people taking this view are guilty of the above confusion. Some really hanker after a purpose of life in this sense. To some people the greatest attraction of the medieval world picture is the belief in an omnipotent, omniscient, and all-good Father, the view of themselves as His children who worship Him, of their proper attitude to what befalls them as submission, humility, resignation in His will, and what is often described as the "creaturely feeling."[16] All these are attitudes and feelings appropriate to a being that stands to another in the same sort of relation, though of course on a higher plane, in which a helpless child stands to his progenitor. Many regard the scientific picture of the world as cold, unsympathetic, unhomely, frightening, because it does not provide for any appropriate object of this creaturely attitude. There is nothing and no one in the world, as science depicts it, in which we can have faith or trust, on whose guidance we can rely, to whom we can turn for consolation, whom we can worship or submit to—except other human beings. This may be felt as a keen disappointment, because it shows that the meaning of life cannot lie in submission to His will, in acceptance of whatever may come, and in worship. But it does not imply that life can have *no* meaning. It merely implies that it must have a different meaning from that which it was thought to have. Just as it is a great shock for a child to find that he must stand on his own feet, that his father and mother no longer provide for him, so a person who has lost his faith in God must reconcile himself to the idea that he has to stand on his own feet, alone in the world except for whatever friends he may succeed in making.

But is not this to miss the point of the Christian teaching? Surely, Christianity can tell us the meaning of life because it tells us the grand and noble end for which God has created the universe and man. No human life, however pointless it may seem, is meaningless because in being part of God's plan, every life is assured of significance.

---

15. See e.g. "Is Life Worth Living?" B.B.C. Talk by the Rev. John Sutherland Bonnell in *Asking Them Questions,* Third Series, ed. by R. S. Wright (London: Geoffrey Cumberlege, 1950).

16. See e.g. Rudolf Otto, *The Idea of the Holy,* pp. 9–11. See also C. A. Campbell, *On Selfhood and Godhood* (London: George Allen & Unwin Ltd., 1957) p. 246, and H. J. Paton, *The Modern Predicament,* pp. 69–71.

This point is well taken. It brings to light a distinction of some importance: we call a person's life meaningful not only if it is worthwhile, but also if he has helped in the realization of some plan or purpose transcending his own concerns. A person who knows he must soon die a painful death, can give significance to the remainder of his doomed life by, say, allowing certain experiments to be performed on him which will be useful in the fight against cancer. In a similar way, only on a much more elevated plane, every man, however humble or plagued by suffering, is guaranteed significance by the knowledge that he is participating in God's purpose.

What, then, on the Christian view, is the grand and noble end for which God has created the world and man in it? We can immediately dismiss that still popular opinion that the smallness of our intellect prevents us from stating meaningfully God's design in all its imposing grandeur.[17] This view cannot possibly be a satisfactory answer to our question about the purpose of life. It is, rather, a confession of the impossibility of giving one. If anyone thinks that this "answer" can remove the sting from the impression of meaninglessness and insignificance in our lives, he cannot have been stung very hard.

If, then, we turn to those who are willing to state God's purpose in so many words, we encounter two insuperable difficulties. The first is to find a purpose grand and noble enough to explain and justify the great amount of undeserved suffering in this world. We are inevitably filled by a sense of bathos when we read statements such as this: ". . . history is the scene of a divine purpose, in which the whole history is included, and Jesus of Nazareth is the centre of that purpose, both as revelation and as achievement, as the fulfilment of all that was past, and the promise of all that was to come. . . . If God is God, and if He made all these things, why did He do it? . . . God created a universe, bounded by the categories

17. For a discussion of this issue, see the eighteenth century controversy between Deists and Theists, for instance, in Sir Leslie Stephen's *History of English Thought in the Eighteenth Century* (London: Smith, Elder & Co., 1902) pp. 112–119 and pp. 134–163. See also the attacks by Toland and Tindal on "the mysterious" in *Christianity not Mysterious* and *Christianity as Old as the Creation, or the Gospel a Republication of the Religion of Nature,* resp., parts of which are reprinted in Henry Bettenson's *Doctrines of the Christian Church,* pp. 426–431. For modern views maintaining that mysteriousness is an essential element in religion, see Rudolf Otto, *The Idea of the Holy,* esp. pp. 25–40, and most recently M. B. Foster, *Mystery and Philosophy* (London: S.C.M. Press, 1957) esp. Chs. IV. and VI. For the view that statements about God must be nonsensical or absurd, see e.g. H. J. Paton, op. cit. pp. 119–120, 367–369. See also "Theology and Falsification" in *New Essays in Philosophical Theology,* ed. by A. Flew and A. MacIntyre (London: S.C.M. Press, 1955) pp. 96–131; also N. McPherson, "Religion as the Inexpressible," ibid., esp. pp. 137–143.

of time, space, matter, and causality, because He desired to enjoy for ever the society of a fellowship of finite and redeemed spirits which have made to His love the response of free and voluntary love and service."[18] Surely this cannot be right? Could a God be called omniscient, omnipotent, *and* all-good who, for the sake of satisfying his desire to be loved and served, imposes (or has to impose) on his creatures the amount of undeserved suffering we find in the world?

There is, however, a much more serious difficulty still: God's purpose in making the universe must be stated in terms of a dramatic story many of whose key incidents symbolize religious conceptions and practices which we no longer find morally acceptable: the imposition of a taboo on the fruits of a certain tree, the sin and guilt incurred by Adam and Eve by violating the taboo, the wrath of God,[19] the curse of Adam and Eve and all their progeny, the expulsion from Paradise, the Atonement by Christ's bloody sacrifice on the cross which makes available by way of the sacraments God's Grace by which alone men can be saved (thereby, incidentally, establishing the valuable power of priests to forgive sins and thus alone make possible a man's entry to heaven,[20]) Judgment Day on which the sheep are separated from the goats and the latter condemned to eternal torment in hell-fire.

Obviously it is much more difficult to formulate a purpose for creating the universe and man that will justify the enormous amount of undeserved suffering which we find around us, if that story has to be fitted in as well. For now we have to explain not only why an omnipotent, omniscient, and all-good God should create such a universe and such a man, but also why, foreseeing every move of the feeble, weak-willed, ignorant, and covetous creature to be created, He should nevertheless have created him and, having done so, should be incensed and outraged by man's sin, and why He should deem it necessary to sacrifice His own son on the cross to atone for this sin which was, after all, only a disobedience of one of his commands, and why this atonement and consequent redemption could not have been followed by man's return to Paradise—particularly of

18. Stephen Neill, *Christian Faith To-day* (London: Penguin Books, 1955) pp. 240–241.

19. It is difficult to feel the magnitude of this first sin unless one takes seriously the words "Behold, the man has eaten of the fruit of the tree of knowledge of good and evil, and is become as one of us; and now, may he not put forth his hand, and take also of the tree of life, and eat, and live for ever?" Genesis iii, 22.

20. See in this connection the pastoral letter of 2nd February, 1905, by Johannes Katschtaler, Prince Bishop of Salzburg on the honour due to priests, contained in *Quellen zur Geschichte des Papsttums*, by Mirbt pp. 497–499, translated and reprinted in *The Protestant Tradition*, by J. S. Whale (Cambridge: University Press, 1955) pp. 259–262.

those innocent children who had not yet sinned—and why, on Judgment Day, this merciful God should condemn some to eternal torment.[21] It is not surprising that in the face of these and other difficulties, we find, again and again, a return to the first view: that God's purpose cannot meaningfully be stated.

It will perhaps be objected that no Christian to-day believes in the dramatic history of the world as I have presented it. But this is not so. It is the official doctrine of the Roman Catholic, the Greek Orthodox, and a large section of the Anglican Church.[22] Nor does Protestantism substantially alter this picture. In fact, by insisting on "Justification by Faith Alone" and by rejecting the ritualistic, magical character of the medieval Catholic interpretation of certain elements in the Christian religion, such as indulgences, the sacraments, and prayer, while at the same time insisting on the necessity of grace, Protestantism undermined the moral element in medieval Christianity expressed in the Catholics' emphasis on personal merit.[23] Protestantism, by harking back to St. Augustine, who clearly realized the incompatibility of grace and personal merit,[24] opened the way for Calvin's doctrine of Predestination (the intellectual parent of that form of rigid determinism which is usually blamed on science) and Salvation or Condemnation from all eternity.[25] Since Roman Catholics, Lutherans, Calvinists, Presbyterians and Baptists officially subscribe to the views just outlined, one can justifiably claim that the overwhelming majority of professing Christians hold or ought to hold them.

It might still be objected that the best and most modern views are wholly different. I have not the necessary knowledge to pronounce on the accuracy of this claim. It may well be true that the best and most modern views are such as Professor Braithwaite's who maintains that Christianity is, roughly speaking, "morality plus stories," where the stories are intended merely to make the strict moral teaching both more easily understandable

---

21. How impossible it is to make sense of this story has been demonstrated beyond any doubt by Tolstoy in his famous "Conclusion of A Criticism of Dogmatic Theology," reprinted in *A Confession, The Gospel in Brief, and What I Believe.*
22. See "The Nicene Creed," "The Tridentine Profession of Faith," "The Syllabus of Errors," reprinted in *Documents of the Christian Church,* pp. 34, 373 and 380 resp.
23. See e.g. J. S. Whale, *The Protestant Tradition,* Ch. IV., esp. pp. 48–56.
24. See ibid., pp. 61 ff.
25. See "The Confession of Augsburg" esp. Articles II., IV., XVIII., XIX., XX.; "Christianae Religionis Institutio," "The Westminster Confession of Faith," esp. Articles III., VI., IX., X., XI., XVI., XVII.; "The Baptist Confession of Faith," esp. Articles III., XXI., XXIII., reprinted in *Documents of the Christian Church,* pp. 294 ff., 298 ff., 344 ff., 349 ff.

and more palatable.[26] Or it may be that one or the other of the modern views on the nature and importance of the dramatic story told in the sacred Scriptures is the best. My reply is that even if it is true, it does not prove what I wish to disprove, that one can extract a sensible answer to our question, "What is the meaning of life?" from the kind of story subscribed to by the overwhelming majority of Christians, who would, moreover, reject any such modernist interpretation at least as indignantly as the scientific account. Moreover, though such views can perhaps avoid some of the worst absurdities of the traditional story, they are hardly in a much better position to state the purpose for which God has created the universe and man in it, because they cannot overcome the difficulty of finding a purpose grand and noble enough to justify the enormous amount of undeserved suffering in the world.

Let us, however, for argument's sake, waive all these objections. There remains one fundamental hurdle which no form of Christianity can overcome: the fact that it demands of man a morally repugnant attitude towards the universe. It is now very widely held[27] that the basic element of the Christian religion is an attitude of worship towards a being supremely worthy of being worshipped and that it is religious feelings and experiences which apprise their owner of such a being and which inspire in him the knowledge or the feeling of complete dependence, awe, worship, mystery, and self-abasement. There is, in other words, a bi-polarity (the famous "I-Thou relationship") in which the object, "the wholly-other," is exalted whereas the subject is abased to the limit. Rudolf Otto has called this the "creature-feeling"[28] and he quotes as an expression of it, Abraham's words when venturing to plead for the men of Sodom: "Behold now, I have taken upon me to speak unto the Lord, which am but dust and ashes." (Gen. XVIII.27). Christianity thus demands of men an attitude inconsistent with one of the presuppositions of morality: that man is not wholly dependent on something else, that man has free will, that man is in principle capable of responsibility. We have seen that the concept of grace is the Christian attempt to reconcile the claim of total dependence and the claim of individual responsibility (partial independence), and it is obvious that such attempts must fail. We may dismiss certain doctrines,

26. See e.g. his *An Empiricist's View of the Nature of Religious Belief* (Eddington Memorial Lecture).
27. See e.g. the two series of Gifford Lectures most recently published: *The Modern Predicament* by H. J. Paton (London: George Allen & Unwin Ltd., 1955) pp. 69 ff., and *On Selfhood and Godhood* by C. A. Campbell (London: George Allen & Unwin Ltd., 1957) pp. 231–250.
28. Rudolf Otto, *The Idea of the Holy*, p. 9.

such as the doctrine of original sin or the doctrine of eternal hellfire or the doctrine that there can be no salvation outside the Church as extravagant and peripheral, but we cannot reject the doctrine of total dependence without rejecting the characteristically Christian attitude as such.

3. THE MEANING OF LIFE

Perhaps some of you will have felt that I have been shirking the real problem. To many people the crux of the matter seems as follows. How can there be any meaning in our life if it ends in death? What meaning can there be in it that our inevitable death does not destroy? How can our existence be meaningful if there is no after-life in which perfect justice is meted out? How can life have any meaning if all it holds out to us are a few miserable earthly pleasures and even these to be enjoyed only rarely and for such a piteously short time?

I believe this is the point which exercises most people most deeply. Kirilov, in Dostoevsky's novel, *The Possessed*, claims, just before committing suicide, that as soon as we realize that there is no God, we cannot live any longer, we must put an end to our lives. One of the reasons which he gives is that when we discover that there is no paradise, we have nothing to live for.

". . . there was a day on earth, and in the middle of the earth were three crosses. One on the cross had such faith that He said to another, 'To-day thou shalt be with me in paradise.' The day came to an end, both died, and they went, but they found neither paradise nor resurrection. The saying did not come true. Listen: that man was the highest of all on earth. . . . There has never been any one like Him before or since, and never will be. . . . And if that is so, if the laws of Nature did not spare even *Him*, and made even Him live in the midst of lies and die for a lie, then the whole planet is a lie and is based on a lie and a stupid mockery. So the very laws of the planet are a lie and a farce of the devil. What, then, is there to live for?"[29] And Tolstoy, too, was nearly driven to suicide when he came to doubt the existence of God and an after-life.[30] And this is true of many.

What, then, is it that inclines us to think that if life is to have a meaning, there would be an after-life? It is this. The Christian world view

29. Fyodor Dostoyevsky, *The Devils* (London: The Penguin Classics, 1953) pp. 613–614.
30. Leo Tolstoy, *A Confession, The Gospel in Brief, and What I Believe,* The World's Classics, p. 24.

contains the following three propositions. The first is that since the Fall, God's curse of Adam and Eve, and the expulsion from Paradise, life on earth for mankind has not been worth while, but a vale of tears, one long chain of misery, suffering, unhappiness, and injustice. The second is that a perfect after-life is awaiting us after the death of the body. The third is that we can enter this perfect life only on certain conditions, among which is also the condition of enduring our earthly existence to its bitter end. In this way, our earthly existence which, in itself, would not (at least for many people if not all) be worth living, acquires meaning and significance: only if we endure it, can we gain admission to the realm of the blessed.

It might be doubted whether this view is still held to-day. However, there can be no doubt that even to-day we all imbibe a good deal of this view with our earliest education. In sermons, the contrast between the perfect life of the blessed and our life of sorrow and drudgery is frequently driven home and we hear it again and again that Christianity has a message of hope and consolation for all those "who are weary and heavy laden."[31]

It is not surprising, then, that when the implications of the scientific world picture begin to sink in, when we come to have doubts about the existence of God and another life, we are bitterly disappointed. For if there is no afterlife, then all we are left is our earthly life which we have come to regard as a necessary evil, the painful fee of admission to the land of eternal bliss. But if there is no eternal bliss to come and if this hell on earth is all, why hang on till the horrible end?

Our disappointment therefore arises out of these two propositions, that the earthly life is not worth living, and that there is another perfect life of eternal happiness and joy which we may enter upon if we satisfy certain conditions. We can regard our lives as meaningful, if we believe both. We cannot regard them as meaningful if we believe merely the first and not the second. It seems to me inevitable that people who are taught something of the history of science, will have serious doubts about the second. If they cannot overcome these, as many will be unable to do, then they must either accept the sad view that their life is meaningless or they must abandon the first proposition: that this earthly life is not worth living. They must find the meaning of their life in this earthly existence. But is this possible?

A moment's examination will show us that the Christian evaluation of our earthly life as worthless, which we accept in our moments of pessi-

31. See for instance J. S. Whale, *Christian Doctrine*, pp. 171, 176–178, etc. See also Stephen Neill, *Christian Faith To-day*, p. 241.

mism and dissatisfaction, is not one that we normally accept. Consider
only the question of murder and suicide. On the Christian view, other
things being equal, the most kindly thing to do would be for every one of
us to kill as many of our friends and dear ones as still have the misfortune
to be alive, and then to commit suicide without delay, for every moment
spent in this life is wasted. On the Christian view, God has not made it
that easy for us. He has forbidden us to hasten others or ourselves into the
next life. Our bodies are his private property and must be allowed to wear
themselves out in the way decided by Him, however painful and horrible
that may be. We are, as it were, driving a burning car. There is only one
way out, to jump clear and let it hurtle to destruction. But the owner of
the car has forbidden it on pain of eternal tortures worse than burning.
And so we do better to burn to death inside.

On this view, murder is a less serious wrong than suicide. For murder
can always be confessed and repented and therefore forgiven, suicide cannot—unless we allow the ingenious way out chosen by the heroine of
Graham Greene's play, The Living Room, who swallows a slow but deadly
poison and, while awaiting its taking effect, repents having taken it. Murder, on the other hand, is not so serious because, in the first place, it need
not rob the victim of anything but the last lap of his march in the vale of
tears, and, in the second place, it can always be forgiven. Hamlet, it will
be remembered, refrains from killing his uncle during the latter's prayers
because, as a true Christian, he believes that killing his uncle at that
point, when the latter has purified his soul by repentance, would merely
be doing him a good turn, for murder at such a time would simply des-
patch him to undeserved and everlasting happiness.

These views strike us as odd, to say the least. They are the logical
consequence of the official medieval evaluation of this our earthly existence. If this life is not worth living, then taking it is not robbing the person concerned of much. The only thing wrong with it is the damage to
God's property, which is the same both in the case of murder and suicide.
We do not take this view at all. Our view, on the contrary, is that murder
is the most serious wrong because it consists in taking away from some
one else against his will his most precious possession, his life. For this reason, when a person suffering from an incurable disease asks to be killed,
the mercy killing of such a person is regarded as a much less serious crime
than murder because, in such a case, the killer is not robbing the other of
a good against his will. Suicide is not regarded as a real crime at all, for
we take the view that a person can do with his own possessions what he
likes.

However, from the fact that these are our normal opinions, we can infer nothing about their truth. After all, we could easily be mistaken. Whether life is or is not worthwhile, is a value judgment. Perhaps all this is merely a matter of opinion or taste. Perhaps no objective answer can be given. Fortunately, we need not enter deeply into these difficult and controversial questions. It is quite easy to show that the medieval evaluation of earthly life is based on a misguided procedure.

Let us remind ourselves briefly of how we arrive at our value judgments. When we determine the merits of students, meals, tennis players, bulls, or bathing belles, we do so on the basis of some criteria and some standard or norm. Criteria and standards notoriously vary from field to field and even from case to case. But that does not mean that we have *no* idea about what are the appropriate criteria or standards to use. It would not be fitting to apply the criteria for judging bulls to the judgment of students or bathing belles. They score on quite different points. And even where the same criteria are appropriate as in the judgment of students enrolled in different schools and universities, the standards will vary from one institution to another. Pupils who would only just pass in one, would perhaps obtain honours in another. The higher the standard applied, the lower the marks, that is, the merit conceded to the candidate.

The same procedure is applicable also in the evaluation of a life. We examine it on the basis of certain criteria and standards. The medieval Christian view uses the criteria of the ordinary man: a life is judged by what the person concerned can get out of it: the balance of happiness over unhappiness, pleasure over pain, bliss over suffering. Our earthly life is judged not worthwhile because it contains much unhappiness, pain, and suffering, little happiness, pleasure, and bliss. The next life is judged worthwhile because it provides eternal bliss and no suffering.

Armed with these criteria, we can compare the life of this man and that, and judge which is more worthwhile, which has a greater balance of bliss over suffering. But criteria alone enable us merely to make comparative judgments of value, not absolute ones. We can say which is more and which is less worthwhile, but we cannot say which is worthwhile and which is not. In order to determine the latter, we must introduce a standard. But what standard ought we to choose?

Ordinarily, the standard we employ is the average of the kind. We call a man and a tree tall if they are well above the average of their kind. We do not say that Jones is a short man because he is shorter than a tree. We do not judge a boy a bad student because his answer to a question in the Leaving Examination is much worse than that given in reply to the same question by a young man sitting for his finals for the Bachelor's degree.

The same principles must apply to judging lives. When we ask whether a given life was or was not worthwhile, then we must take into consideration the range of worthwhileness which ordinary lives normally cover. Our end poles of the scale must be the best possible and the worst possible life that one finds. A good and worthwhile life is one that is well above average. A bad one is one well below.

The Christian evaluation of earthly lives is misguided because it adopts a quite unjustifiably high standard. Christianity singles out the major shortcomings of our earthly existence: there is not enough happiness; there is too much suffering; the good and bad points are quite unequally and unfairly distributed; the underprivileged and underendowed do not get adequate compensation; it lasts only a short time. It then quite accurately depicts the perfect or ideal life as that which does not have any of these shortcomings. Its next step is to promise the believer that he will be able to enjoy this perfect life later on. And then it adopts as its standard of judgment the perfect life, dismissing as inadequate anything that falls short of it. Having dismissed earthly life as miserable, it further damns it by characterizing most of the pleasures of which earthly existence allows as bestial, gross, vile, and sinful, or alternatively as not really pleasurable.

This procedure is as illegitimate as if I were to refuse to call anything tall unless it is infinitely tall, or anything beautiful unless it is perfectly flawless, or any one strong unless he is omnipotent. Even if it were true that there is available to us an after-life which is flawless and perfect, it would still not be legitimate to judge earthly lives by this standard. We do not fail every candidate who is not an Einstein. And if we do not believe in an after-life, we must of course use ordinary earthly standards.

I have so far only spoken of the worthwhileness, only of what a person can get out of a life. There are other kinds of appraisal. Clearly, we evaluate people's lives not merely from the point of view of what they yield to the persons that lead them, but also from that of other men on whom these lives have impinged. We judge a life more significant if the person has contributed to the happiness of others, whether directly by what he did for others, or by the plans, discoveries, inventions, and work he performed. Many lives that hold little in the way of pleasure or happiness for its owner are highly significant and valuable, deserve admiration and respect on account of the contributions made.

It is now quite clear that death is simply irrelevant. If life can be worthwhile at all, then it can be so even though it be short. And if it is not worthwhile at all, then an eternity of it is simply a nightmare. It may be sad that we have to leave this beautiful world, but it is so only if and because it is beautiful. And it is no less beautiful for coming to an end. I rather suspect that an eternity of it might make us less appreciative, and in the end it would be tedious.

It will perhaps be objected now that I have not really demonstrated that life has a meaning, but merely that it can be worthwhile or have value. It must be admitted that there is a perfectly natural interpretation of the question, "What is the meaning of life?" on which my view actually proves that life has no meaning. I mean the interpretation discussed in section 2 of this lecture, where I attempted to show that, if we accept the explanations of natural science, we cannot believe that living organisms have appeared on earth in accordance with the deliberate plan of some intelligent being. Hence, on this view, life cannot be said to have a purpose, in the sense in which man-made things have a purpose. Hence it cannot be said to have a meaning or significance in that sense.

However, this conclusion is innocuous. People are disconcerted by the thought that *life as such* has no meaning in that sense only because they very naturally think it entails that no individual life can have meaning either. They naturally assume that *this* life or *that* can have meaning only if *life as such* has meaning. But it should by now be clear that your life and mine may or may not have meaning (in one sense) even if life as such has none (in the other). Of course, it follows from this that your life may have meaning while mine has not. The Christian view guarantees a meaning (in one sense) to every life, the scientific view does not (in any sense). By relating the question of the meaningfulness of life to the particular circumstances of an individual's existence, the scientific view leaves it an open question whether an individual's life has meaning or not. It is, however, clear that the latter is the important sense of "having a meaning." Christians, too, must feel that their life is wasted and meaningless if they have not achieved salvation. To know that even such lost lives have a meaning in another sense is no consolation to them. What matters is not that life should have a guaranteed meaning, whatever happens here or here-after, but that, by luck (Grace) or the right temperament and attitude (Faith) or a judicious life (Works) a person should make the most of his life.

"But here lies the rub," it will be said. "Surely, it makes all the difference whether there is an after-life. This is where morality comes in." It would be a mistake to believe that. Morality is not the meting out of punishment and reward. To be moral is to refrain from doing to others what, if they followed reason, they would not do to themselves, and to do for others what, if they followed reason, they would want to have done. It is, roughly speaking, to recognize that others, too, have a right to a worthwhile life. Being moral does not make one's own life worthwhile, it helps others to make theirs so.

CONCLUSION

I have tried to establish three points: (i) that scientific explanations render their explicanda as intelligible as pre-scientific explanations; they differ from the latter only in that, having testable implications and being more precisely formulated, their truth or falsity can be determined with a high degree of probability; (ii) that science does not rob human life of purpose, in the only sense that matters, but, on the contrary, renders many more of our purposes capable of realization; (iii) that common sense, the Christian world view, and the scientific approach agree on the criteria but differ on the standard to be employed in the evaluation of human lives; judging human lives by the standards of perfection, as Christians do, is unjustified; if we abandon this excessively high standard and replace it by an everyday one, we have no longer any reason for dismissing earthly existence as not worthwhile.

On the basis of these three points I have attempted to explain why so many people come to the conclusion that human existence is meaningless and to show that this conclusion is false. In my opinion, this pessimism rests on a combination of two beliefs, both partly true and partly false: the belief that the meaningfulness of life depends on the satisfaction of at least three conditions, and the belief that this universe satisfies none of them. The conditions are, first, that the universe is intelligible, second, that life has a purpose, and third, that all men's hopes and desires can ultimately be satisfied. It seemed to medieval Christians and it seems to many Christians to-day that Christianity offers a picture of the world which can meet these conditions. To many Christians and non-Christians alike it seems that the scientific world picture is incompatible with that of Christianity, therefore with the view that these three conditions are met, therefore with the view that life has a meaning. Hence they feel that they are confronted by the dilemma of accepting either a world picture incompatible with the discoveries of science or the view that life is meaningless.

I have attempted to show that the dilemma is unreal because life can be meaningful even if not all of these conditions are met. My main conclusion, therefore, is that acceptance of the scientific world picture provides no reason for saying that life is meaningless, but on the contrary every reason for saying that there are many lives which are meaningful and significant. My subsidiary conclusion is that one of the reasons frequently offered for retaining the Christian world picture, namely, that its acceptance gives us a guarantee of a meaning for human existence, is unsound. We can see that our lives can have a meaning even if we abandon it and adopt the scientific world picture instead. I have, moreover, mentioned several reasons for rejecting the Christian world picture: (i) the biblical explanations of the details of our universe are often simply false; (ii) the so-called explanations of the whole universe are incomprehensible or absurd; (iii) Christianity's low evaluation of earthly existence (which is the main cause of the belief in the meaninglessness of life) rests on the use of an unjustifiably high standard of judgment.

❧ 9 ❧

PAUL EDWARDS

The Meaning
and Value of Life

To the questions "Is human life ever worthwhile?" and "Does (or can) human life have any meaning?" many religious thinkers have offered affirmative answers with the proviso that these answers would not be justified unless two of the basic propositions of most Western religions were true—that human life is part of a divinely ordained cosmic scheme and that after death at least some human beings will be rewarded with eternal bliss. Thus, commenting on Bertrand Russell's statement that not only must each individual human life come to an end but that life in general will eventually die out, C. H. D. Clark contrasts this "doctrine of despair" with the beauty of the Christian scheme. "If we are asked to believe that all our striving is without final consequence," then "life is meaningless and it scarcely matters how we live if all will end in the dust of death." According to Christianity, on the other hand, "each action has vital significance." Clark assures us that "God's grand design is life eternal for those who walk in the steps of Christ. Here is the one grand incentive to good living. . . . As life is seen to have purpose and meaning, men find release from despair and the fear of death" (*Christianity and Bertrand Russell*, p. 30). In a similar vein, the Jewish existentialist Emil Fackenheim claims that "whatever meaning life acquires" is derived from the encounter between God and man. The meaning thus conferred upon human life "cannot be understood in terms of some finite human purpose, supposedly more ultimate than the meeting itself. For what could be more ultimate than the Presence of God?" It is true that God is not always "near," but "times of Divine farness" are by no means devoid of meaning. "Times of Divine nearness do not light up themselves alone.

Their meaning extends over all of life." There is a "dialectic between Divine nearness and Divine farness," and it points to "an eschatological future in which it is overcome" ("Judaism and the Meaning of Life").

Among unbelievers not a few maintain that life can be worthwhile and have meaning in some humanly important sense even if the religious world view is rejected. Others, however, agree with the religious theorists that our two questions must be given negative answers if there is no God and if death means personal annihilation. Having rejected the claims of religion, they therefore conclude that life is not worthwhile and that it is devoid of meaning. These writers, to whom we shall refer here as "pessimists," do not present their judgments as being merely expressions of certain moods or feelings but as conclusions that are in some sense objectively warranted. They offer reasons for their conclusions and imply that anybody reaching a contradictory conclusion is mistaken or irrational. Most pessimists do not make any clear separation between the statements that life is not worthwhile and that life is without meaning. They usually speak of the "futility" or the "vanity" of life, and presumably they mean by this both that life is not worth living and that it has no meaning. For the time being we, too, shall treat these statements as if they were equivalent. However, later we shall see that in certain contexts it becomes important to distinguish between them.

Our main concern in this article will be to appraise pessimism as just defined. We shall not discuss either the question whether life is part of a divinely ordained plan or the question whether we survive our bodily death. Our question will be whether the pessimistic conclusions are justified if belief in God and immortality are rejected.

SCHOPENHAUER'S ARGUMENTS

Let us begin with a study of the arguments offered by the pessimists, remembering that many of these are indirectly endorsed by religious apologists. The most systematic and probably the most influential, though in fact not the gloomiest, of the pessimists was Schopenhauer. The world, he wrote, is something which ought not to exist: the truth is that "we have not to rejoice but rather to mourn at the existence of the world; that its non-existence would be preferable to its existence; that it is something which ought not to be." It is absurd to speak of life as a gift, as so many philosophers and thoughtless people have done. "It is evident that everyone would have declined such a gift if he could have seen it and tested it beforehand." To those who assure us that life is only a lesson, we are entitled to reply: "For this very reason I wish I had been left in the peace of the all-sufficient nothing, where I would have no need of lessons or of anything else" (*The World as Will and Idea*, Vol. III, p. 390).

Schopenhaeur offers numerous arguments for his conclusion. Some of these are purely metaphysical and are based on his particular system. Others, however, are of a more empirical character and are logically independent of his brand of metaphysical voluntarism. Happiness, according to Schopenhauer, is unobtainable for the vast majority of mankind. "Everything in life shows that earthly happiness is destined to be frustrated or recognized as illusion." People either fail to achieve the ends they are striving for or else they do achieve them only to find them grossly disappointing. But as soon as a man discovers that a particular goal was not really worth pursuing, his eye is set on a new one and the same illusory quest begins all over again. Happiness, accordingly, always lies in the future or in the past, and "the present may be compared to a small dark cloud which the wind drives over the sunny plain: before and behind it all is bright, only it itself always casts a shadow. The present is therefore always insufficient; but the future is uncertain, and the past is irrevocable" (ibid., p. 383). Men in general, except for those sufficiently rational to become totally resigned, are constantly deluded—"now by hope, now by what was hoped for." They are taken in by "the enchantment of distance," which shows them "paradises." These paradises, however, vanish like "optical illusions when we have allowed ourselves to be mocked by them." The "fearful envy" excited in most men by the thought that somebody else is genuinely happy shows how unhappy they really are, whatever they pretend to others or to themselves. It is only "because they feel themselves unhappy" that "men cannot endure the sight of one whom they imagine happy."

On occasions Schopenhauer is ready to concede that some few human beings really do achieve "comparative" happiness, but this is not of any great consequence. For aside from being "rare exceptions," these happy people are really like "decoy birds"—they represent a possibility which must exist in order to lure the rest of mankind into a false sense of hope. Moreover, happiness, insofar as it exists at all, is a purely "negative" reality. We do not become aware of the greatest blessings of life—health, youth, and freedom—until we have lost them. What is called pleasure or satisfaction is merely the absence of craving or pain. But craving and pain are positive. As for the few happy days of our life—if there are any—we notice them only "after they have given place to unhappy ones."

Schopenhauer not infrequently lapsed from his doctrine of the "negative" nature of happiness and pleasure into the more common view that their status is just as "positive" as that of unhappiness and pain. But he had additional arguments which do not in any way depend on the theory that happiness and pleasure are negative. Perhaps the most important of these is the argument from the "perishableness" of all good things and the ultimate extinction of all our hopes and achievements in death. All our pleasures and joys "disappear in our hands, and we afterwards ask astonished where they have gone." Moreover, a joy which no longer exists does not "count"—it counts as little as if it had never been experienced at all:

> That which *has been* exists no more; it exists as little as that which has *never* been. But of everything that exists you may say, in the next moment, that it has been. Hence something of great importance in our past is inferior to something of little importance in our present, in that the latter is a *reality,* and related to the former as something to nothing. ("The Vanity of Existence," in *The Will to Live,* p. 229)

Some people have inferred from this that the enjoyment of the present should be "the supreme object of life." This is fallacious; for "that which in the next moment exists no more, and vanishes utterly, like a dream, can never be worth a serious effort."

The final "judgment of nature" is destruction by death. This is "the last proof" that life is a "false path," that all man's wishing is "a perversity," and that "nothing at all is worth our striving, our efforts and struggles." The conclusion is inescapable: "All good things are vanity, the world in all its ends bankrupt, and life a business which does not cover its expenses" (*The World as Will and Idea,* Vol. III, p. 383).

THE POINTLESSNESS OF IT ALL

Some of Schopenhauer's arguments can probably be dismissed as the fantasies of a lonely and embittered man who was filled with contempt for mankind and who was singularly incapable of either love or friendship. His own misery, it may be plausibly said, made Schopenhauer overestimate the unhappiness of human beings. It is frequently, but not universally, true that what is hoped for is found disappointing when it is attained, and while "fearful envy" of other people's successes is common enough, real sympathy and generosity are not quite so rare as Schopenhauer made them out to be. Furthermore, his doctrine that pleasure is negative while pain is positive, insofar as one can attach any clear meaning to it, seems glaringly false. To this it should be added, however, that some of Schopenhauer's arguments are far from idiosyncratic and that substantially the same conclusions have been endorsed by men who were neither lonely nor embittered and who did not, as far as one can judge, lack the gift of love or friendship.

Clarence Darrow, one of the most compassionate men who ever lived, also concluded that life was an "awful joke." Like Schopenhauer, Darrow offered as one of his reasons the apparent aimlessness of all that happens. "This weary old world goes on, begetting, with birth and with living and with death," he remarked in his moving plea for the boy-murderers Loeb and Leopold, "and all of it is blind from the beginning to the end" (*Clarence Darrow—Attorney for the Damned*, A. Weinberg, ed., New York, 1957). Elsewhere he wrote: "Life is like a ship on the sea, tossed by every wave and by every wind; a ship headed for no port and no harbor, with no rudder, no compass, no pilot; simply floating for a time, then lost in the waves" ("Is Life Worth Living?," p. 43). In addition to the aimlessness of life and the universe, there is the fact of death. "I love my friends," wrote Darrow, "but they all must come to a tragic end." Death is more terrible the more one is attached to things in the world. Life, he concludes, is "not worth while," and he adds (somewhat inconsistently, in view of what he had said earlier) that "it is an unpleasant interruption of nothing, and the best thing you can say of it is that it does not last long" ("Is the Human Race Getting Anywhere?," p. 53).

Tolstoy, unlike Darrow, eventually came to believe in Christianity, or at least in his own idiosyncratic version of Christianity, but for a number of years the only position for which he could see any rational justification was an extreme form of pessimism. During that period (and there is reason to believe that in spite of his later protestations to the contrary, his feelings on this subject never basically changed) Tolstoy was utterly overwhelmed by the thought of his death and the death of those he cared for and, generally, by the transitory nature of all human achievements. "Today or tomorrow," he wrote in "A Confession," "sickness and death will come to those I love or to me; nothing will remain but stench and worms. Sooner or later my affairs, whatever they may be, will be forgotten, and I shall not exist. Then why go on making any effort?" Tolstoy likened the fate of man to that of the traveler in the Eastern tale who, pursued by an enraged beast, seeks refuge in a dry well. At the bottom of the well he sees a dragon that has opened its jaws to swallow him. To escape the enraged beast above and the dragon below, he holds onto a twig that is growing in a crack in the well. As he looks around he notices that two mice are gnawing at the stem of the twig. He realizes that very soon the twig will snap and he will fall to his doom, but at the same time he sees some drops of honey on the leaves of the branch and reaches out with his tongue to lick them. "So I too clung to the twig of life, knowing that the dragon of death was inevitably awaiting me, ready to tear me to pieces. . . . I tried to lick the honey which formerly consoled me, but the honey no longer gave me pleasure. . . . I only saw the unescapable dragon and the mice, and I could not tear my gaze from them. And this is not a fable but the real unanswerable truth."

These considerations, according to Tolstoy, inevitably lead to the conclusion that life is a "stupid fraud," that no "reasonable meaning" can be given to a single action or to a whole life. To the questions "What is it for?" "What then?" "Why should I live?" the answer is "Nothing can come of it," "Nothing is worth doing," "Life is not worthwhile."

What ways out are available to a human being who finds himself in this "terrible position"? Judging by the conduct of the people he observed, Tolstoy relates that he could see only four possible "solutions." The first is the way of ignorance. People who adopt this solution (chiefly women and very young and very dull people) have simply not or not yet faced the questions that were tormenting him. Once a person has fully realized what death means, this solution is not available to him. The second way is that of "Epicureanism," which consists in admitting the "hopelessness of life" but seizing as many of life's pleasures as possible while they are within reach. It consists in "disregarding the dragon and the mice and licking the honey in the best way, especially if much of it is around." This, Tolstoy adds, is the solution adopted by the majority of the people belonging to his "circle," by which he presumably means the well-to-do intellectuals of his day. Tolstoy rejects this solution because the vast majority of human beings are not well-to-do and hence have little or no honey at their disposal and also because it is a matter of accident whether one is among those who have honey or those who have not. Moreover, Tolstoy observes, it requires a special "moral dullness," which he himself lacked, to enjoy the honey while knowing the truth about death and the deprivations of the great majority of men. The third solution is suicide. Tolstoy calls this the way of "strength and energy." It is chosen by a few "exceptionally strong and consistent people." After they realize that "it is better to be dead than to be alive, and that it is best of all not to exist," they promptly end the whole "stupid joke." The means for ending it are readily at hand for everybody, but most people are too cowardly or too irrational to avail themselves of them. Finally, there is the way of "weakness." This consists in seeing the dreadful truth and clinging to life never-theless. People of this kind lack the strength to act rationally and Tolstoy adds that he belonged to this last category.

Is it possible for somebody who shares the pessimists' rejection of religion to reach different conclusions without being plainly irrational? Whatever reply may be possible, any intelligent and realistic person would surely have to concede that there is much truth in the pessimists' claims. That few people achieve real and lasting happiness, that the joys of life (where there are any) pass away much too soon, that totally unpredictable events frequently upset the best intentions and wreck the noblest plans—this and much more along the same lines is surely undeniable. Although one should not dogmatize that there will be no significant improvements in the future, the fate of past revolutions, undertaken to rid man of some of his apparently avoidable suffering, does not inspire great hope. The thought of death, too, even in those who are not so overwhelmed by it as Tolstoy, can be quite unendurable. Moreover, to many who have reflected on the implications of physical theory it seems plain that because of the constant increase of entropy in the universe all life anywhere will eventually die out. Forebodings of this kind moved Bertrand Russell to write his famous essay "A Free Man's Worship," in which he concluded that "all the la-bors of the ages, all the devotion, all the inspiration, all the noonday brightness of human genius, are destined to extinction in the vast death of the solar system, and the whole temple of man's achievement must in-evitably be buried beneath the debris of a universe in ruins." Similarly, Wilhelm Ostwald observed that "in the longest run the sum of all human endeavor has no recognizable significance." Although it is disputed whether physical theory really has such gloomy implications, it would perhaps be wisest to assume that the position endorsed by Russell and Ostwald is well-founded.

COMPARATIVE VALUE JUDGMENTS ABOUT LIFE AND DEATH

Granting the strong points in the pessimists' claims, it is still possible to detect certain confusions and dubious inferences in their arguments. To begin with, there is a very obvious inconsistency in the way writers like Darrow and Tolstoy arrive at the conclusion that death is better than life. They begin by telling us that death is something terrible because it terminates the possibility of any of the experiences we value. From this they infer that nothing is really worth doing and that death is better than life. Ignoring for the moment the claim that in view of our inevitable death nothing is "worth doing," there very plainly seems to be an inconsistency in first judging death to be such a horrible evil and in asserting later on that death is better than life. Why was death originally judged to be an evil? Surely because it is the termination of life. And if something, y, is bad because it is the termination of something, x, this can be so only if x is good or has positive value. If x were not good, the termination of x would not be bad. One cannot consistently have it both ways.

To this it may be answered that life did have positive value prior to one's realization of death but that once a person has become aware of the inevitability of his destruction life becomes unbearable and that this is the real issue. This point of view is well expressed in the following exchange between Cassius and Brutus in Shakespeare's *Julius Caesar* (III.i.102–105):

CASSIUS. Why he that cuts off twenty years of life
          Cuts off so many years of fearing death.
BRUTUS.   Grant that, and then is death a benefit:
          So are we Caesar's friends that have abridged
          His time of fearing death.

There is a very simple reply to this argument. Granting that some people after once realizing their doom cannot banish the thought of it from their minds, so much so that it interferes with all their other activities, this is neither inevitable nor at all common. It is, on the contrary, in the opinion of all except some existentialists, morbid and pathological. The realization that one will die does not in the case of most people prevent them from engaging in activities which they regard as valuable or from enjoying the things they used to enjoy. To be told that one is not living "authentically" if one does not brood about death day and night is simply to be insulted gratuitously. A person who knows that his talents are not as great as he would wish or that he is not as handsome as he would have liked to be is not usually judged to live "inauthentically," but on the contrary to be sensible if he does not constantly brood about his limitations and shortcomings and uses whatever talents he does possess to maximum advantage.

There is another and more basic objection to the claim that death is better than life. This objection applies equally to the claim that while death is better than life it would be better still not to have been born in the first place and to the judgment that life is better than death. It should be remembered that we are here concerned with such pronouncements when they are intended not merely as the expression of certain moods but as statements which are in some sense true or objectively warranted. It may be argued that a value comparison—any judgment to the effect that A is better or worse than B or as good as B—makes sense only if *both* A and B are, in the relevant respect, in principle open to inspection. If somebody says, for example, that Elizabeth Taylor is a better actress than Betty Grable, this seems quite intelligible. Or, again, if it is said that life for the Jews is better in the United States than it was in Germany under the Nazis, this also seems readily intelligible. In such cases the terms of the comparison are observable or at any rate describable. These conditions are fulfilled in some cases when value comparisons are made between life and death, but they are not fulfilled in the kind of case with which Tolstoy and the pessimists are concerned. If the conception of an afterlife is intelligible, then it would make sense for a believer or for somebody who has not made up his mind to say such things as "Death cannot be worse than this life" or "I wonder if it will be any better for me after I am dead." Achilles, in the *Iliad*, was not making a senseless comparison when he exclaimed that he would rather act

> . . . as a serf of another,
> A man of little possessions, with scanty means of
>   subsistence,
> Than rule as a ghostly monarch the ghosts of all
>   the departed.

Again, the survivors can meaningfully say about a deceased individual "It is better (for the world) that he is dead" or the opposite. For the person himself, however, if there is no afterlife, death is not a possible object of observation or experience, and statements by him that his own life is better than, as good as, or worse than his own death, unless they are intended to be no more than expressions of certain wishes or moods, must be dismissed as senseless. At first sight the contention that in the circumstances under discussion value comparisons between life and death are senseless may seem implausible because of the widespread tendency to think of death as a shadowy kind of life—as sleep, rest, or some kind of homecoming. Such "descriptions" may be admirable as poetry or consolation, but taken literally they are simply false.

IRRELEVANCE OF THE DISTANT FUTURE

These considerations do not, however, carry us very far. They do not show either that life is worth living or that it "has meaning." Before tackling these problems directly, something should perhaps be said about the curious and totally arbitrary preference of the future to the present, to which writers like Tolstoy and Darrow are committed without realizing it. Darrow implies that life would not be "futile" if it were not an endless cycle of the same kind of activities and if instead it were like a journey toward a destination. Tolstoy clearly implies that life would be worthwhile, that some of our actions at least would have a "reasonable meaning," if the present life were followed by eternal bliss. Presumably, what would make life no longer futile as far as Darrow is concerned is some feature of the destination, not merely the fact that it is a destination; and what would make life worthwhile in Tolstoy's opinion is not merely the eternity of the next life but the "bliss" which it would confer—eternal misery and torture would hardly do. About the bliss in the next life, if there is such a next life, Tolstoy shows no inclination to ask "What for?" or "So what?" But if bliss in the next life is not in need of any further justification, why should any bliss that there might be in the present life need justification?

THE LOGIC OF VALUE JUDGMENTS

Many of the pessimists appear to be confused about the logic of value judgments. It makes sense for a person to ask about something "Is it really worthwhile?" or "Is it really worth the trouble?" if he does not regard it as intrinsically valuable or if he is weighing it against another good with which it may be in conflict. It does not make sense to ask such a question about something he regards as valuable in its own right and where there is no conflict with the attainment of any other good. (This observation, it should be noted, is quite independent of what view one takes of the logical status of intrinsic value judgments.) A person driving to the beach on a crowded Sunday, may, upon finally getting there, reflect on whether the trip was really worthwhile. Or, after undertaking a series of medical treatments, somebody may ask whether it was worth the time and the money involved. Such questions make sense because the discomforts of a car ride and the time and money spent on medical treatments are not usually judged to be valuable for their own sake. Again, a woman who has given up a career as a physician in order to raise a family may ask herself whether it was worthwhile, and in this case the question would make sense not because she regards the raising of a family as no more than a means, but because she is weighing it against another good. However, if somebody is very happy, for any number of reasons—because he is in love, because he won the Nobel prize, because his child recovered from a serious illness—and if this happiness does not prevent him from doing or experiencing anything else he regards as valuable, it would not occur to him to ask "Is it worthwhile?" Indeed, this question would be incomprehensible to him, just as Tolstoy himself would presumably not have known what to make of the question had it been raised about the bliss in the hereafter.

It is worth recalling here that we live not in the distant future but in the present and also, in a sense, in the relatively near future. To bring the subject down to earth, let us consider some everyday occurrences: A man with a toothache goes to a dentist, and the dentist helps him so that the toothache disappears. A man is falsely accused of a crime and is faced with the possibility of a severe sentence as well as with the loss of his reputation; with the help of a devoted attorney his innocence is established, and he is acquitted. It is true that a hundred years later all of the participants in these events will be dead and none of them will *then* be able to enjoy the fruits of any of the efforts involved. But this most emphatically does not imply that the dentist's efforts were not worthwhile or that the attorney's work was not worth doing. To bring in considerations of what will or will not happen in the remote future is, in such and many other though certainly not in all human situations, totally irrelevant. Not only is the finality of death irrelevant here; equally irrelevant are the facts, if they are facts, that life is an endless cycle of the same kind of activities and that the history of the universe is not a drama with a happy ending.

This is, incidentally, also the answer to religious apologists like C. H. D. Clark who maintain that all striving is pointless if it is "without final consequence" and that "it scarcely matters how we live if all will end in the dust of death." Striving is not pointless if it achieves what it is intended to achieve even if it is without *final* consequence, and it matters a great deal how we live if we have certain standards and goals, although we cannot avoid "the dust of death."

THE VANISHED PAST

In asserting the worthlessness of life Schopenhauer remarked that "what has been exists as little as what has never been" and that "something of great importance now past is inferior to something of little importance now present." Several comments are in order here. To begin with, if Schopenhauer is right, it must work both ways: if only the present counts, then past sorrows no less than past pleasures do not "count." Furthermore, the question whether "something of great importance now past is inferior to something of little importance now present" is not, as Schopenhauer supposed, a straightforward question of fact but rather one of valuation, and different answers, none of which can be said to be mistaken, will be given by different people according to their circumstances and interests. Viktor Frankl, the founder of "logotherapy," has compared the pessimist to a man who observes, with fear and sadness, how his wall calendar grows thinner and thinner as he removes a sheet from it every day. The kind of person whom Frankl admires, on the other hand, "files each successive leaf neatly away with its predecessors" and reflects "with pride and joy" on all the richness represented by the leaves removed from the calendar. Such a person will not in old age envy the young. " 'No, thank you,' he will think. 'Instead of possibilities, I have realities in my past' " (*Man's Search for Meaning*, pp. 192–193). This passage is quoted not because it contains any great wisdom but because it illustrates that we are concerned here not with judgments of fact but with value judgments and that Schopenhauer's is not the only one that is possible. Nevertheless, his remarks are, perhaps, a healthy antidote to the cheap consolation and the attempts to cover up deep and inevitable misery that are the stock in trade of a great deal of popular psychology. Although Schopenhauer's judgments about the inferior value of the past cannot be treated as objectively true propositions, they express only too well what a great many human beings are bound to feel on certain occasions. To a man dying of cancer it is small consolation to reflect that there was a time when he was happy and flourishing; and while there are undoubtedly some old people who do not envy the young, it may be suspected that more often the kind of talk advocated by the prophets of positive thinking is a mask for envy and a defense against exceedingly painful feelings of regret and helplessness in the face of aging and death and the now-unalterable past.

THE MEANINGS OF THE "MEANING OF LIFE"

Let us now turn to the question whether, given the rejection of belief in God and immortality, life can nevertheless have any "meaning" or "significance." Kurt Baier has called attention to two very different senses in which people use these expressions and to the confusions that result when they are not kept apart. Sometimes when a person asks whether life has any meaning, what he wants to know is whether there is a superhuman intelligence that fashioned human beings along with other objects in the world to serve some end—whether their role is perhaps analogous to the part of an instrument (or its player) in a symphony. People who ask whether history has a meaning often use the word in the same sense. When Macbeth exclaimed that life "is a tale / Told by an idiot, full of sound and fury, / Signifying nothing," he was answering this cosmic question in the negative. His point evidently was not that human life is part of a scheme designed by a superhuman idiot but that it is not part of any design. Similarly, when Fred Hoyle, in his book *The Nature of the Universe* (rev. ed., New York, 1960), turns to what he calls "the deeper issues" and remarks that we find ourselves in a "dreadful situation" in which there is "scarcely a clue as to whether our existence has any real significance," he is using the word "significance" in this cosmic sense.

On the other hand, when we ask whether a *particular* person's life has or had any meaning, we are usually concerned not with cosmic issues but with the question whether certain purposes are to be found *in* his life. Thus, most of us would say without hesitation that a person's life had meaning if we knew that he devoted himself to a cause (such as the spread of Christianity or communism or the reform of mental institutions), or we would at least be ready to say that it "acquired meaning" once he became sufficiently attached to his cause. Whether we approve of what they did or not, most of us would be ready to admit—to take some random examples—that Dorothea Dix, Pasteur, Lenin, Margaret Sanger, Anthony Comstock, and Winston Churchill led meaningful lives. We seem to mean two things in characterizing such lives as meaningful: we assert, first, that the life in question had some dominant, over-all goal or goals which gave direction to a great many of the individual's actions and, second, that these actions and possibly others not immediately related to the overriding goal were performed with a special zest that was not present before the person became attached to his goal or that would not have been present if there had been no such goal in his life. It is not necessary, however, that a person should be devoted to a cause, in the sense just indicated, before we call his life meaningful. It is sufficient that he should have some attachments that are not too shallow. This last expression is of course rather vague, but so is the use of the word "meaning" when applied to human lives. Since the depth or shallowness of an attachment is a matter of degree, it makes perfectly good sense to speak of degrees of meaning in this context. Thus, C. G. Jung writes that in the lives of his patients there never was "sufficient meaning" (*Memories, Dreams, Reflections*, New York and Toronto, 1963, p. 140). There is nothing odd in such a locution, and there is equally nothing odd in saying about a man who has made a partial recovery from a deep depression that there is now again "some" meaning in his life.

Although frequently when people say about somebody that his life has or had meaning, they evidently regard this as a good thing, this is not invariably the case. One might express this point in the following way: saying that attachment to a certain goal has made a man's life meaningful is *not* tantamount to saying that the acts to which the goal has given direction are of positive value. A man might himself observe—and there would be nothing logically odd about it—"As long as I was a convinced Nazi (or communist or Christian or whatever) my life had meaning, my acts had a zest with which I have not been able to invest them since, and yet most of my actions were extremely harmful." Even while fully devoted to his cause or goal the person need not, and frequently does not, regard it as *intrinsically* valuable. If challenged he will usually justify the attachment to his goal by reference to more fundamental value judgments. Thus, somebody devoted to communism or to medical research or to the dissemination of birth-control information will in all likelihood justify his devotion in terms of the production of happiness and the reduction of suffering, and somebody devoted to Christianity will probably justify his devotion by reference to the will of God.

Let us refer to the first of the two senses we have been discussing as the "cosmic" sense and to the second as the "terrestrial" sense. (These are by no means the only senses in which philosophers and others have used the word "meaning" when they have spoken of the meaning or meaninglessness of life, but for our purposes it is sufficient to take account of these two senses.) Now if the theory of cosmic design is rejected it immediately follows that human life has no meaning in the first or cosmic sense. It does not follow in the least, however, that a particular human life is meaningless in the second, or terrestrial, sense. This conclusion has been very clearly summarized by Baier: "Your life or mine may or may not have meaning (in one sense)," he writes, "even if life as such has none (in the other). . . . The Christian view guarantees a meaning (in one sense) to every life, the scientific view [what we have simply been calling the unbeliever's position] does not in any sense" (*The Meaning of Life*, p. 28). In the terrestrial sense it will be an open question whether an individual's life has meaning or not, to be decided by the particular circumstances of his existence. It may indeed be the case that once a person comes to believe that life has no meaning in the cosmic sense his attachment to terrestrial goals will be undermined to such an extent that his life will cease to be meaningful in the other sense as well. However, it seems very plain that this is by no means what invariably happens, and even if it did invariably happen the meaninglessness of a given person's life in the terrestrial sense would not *logically* follow from the fact, if it is a fact, that life is meaningless in the cosmic sense.

This is perhaps the place to add a few words of protest against the rhetorical exaggerations of certain theological writers. Fackenheim's statement, quoted earlier, that "whatever meaning life acquires, it derives from the encounter between God and man" is typical of many theological pronouncements. Statements of this kind are objectionable on several grounds.

Let us assume that there is a God and that meetings between God and certain human beings do take place; let us also grant that activities commanded by God in these meetings "acquire meaning" by being or becoming means to the end of pleasing or obeying God. Granting all this, it does not follow that obedience of God is the only possible unifying goal. It would be preposterous to maintain that the lives of *all* unbelievers have been lacking in such goals and almost as preposterous to maintain that the lives of believers never contain unifying goals other than obedience of God. There have been devout men who were also attached to the advance of science, to the practice of medicine, or to social reform and who regarded these ends as worth pursuing independently of any divine commandments. Furthermore, there is really no good reason to grant that the life of a particular person becomes meaningful in the terrestrial sense just because human life in general has meaning in the cosmic sense. If a superhuman being has a plan in which I am included, this fact will make (or help to make) my life meaningful in the terrestrial sense only if I know the plan and approve of it and of my place in it, so that working toward the realization of the plan gives direction to my actions.

IS HUMAN LIFE EVER WORTHWHILE?

Let us now turn to the question of whether life is ever worth living. This also appears to be denied by the pessimists when they speak of the vanity or the futility of human life. We shall see that in a sense it cannot be established that the pessimists are "mistaken," but it is also quite easy to show that in at least two senses which seem to be of importance to many people, human lives frequently are worth living. To this end, let us consider under what circumstances a person is likely to raise the question "Is my life (still) worthwhile?" and what is liable to provoke somebody into making a statement like "My life has ceased to be worth living." We saw in an earlier section that when we say of certain acts, such as the efforts of a dentist or a lawyer, that they were worthwhile we are claiming that they achieved certain goals. Something similar seems to be involved when we say that a person's life is (still) worthwhile or worth living. We seem to be making two assertions: first, that the person has some goals (other than merely to be dead or to have his pains eased) which do not seem to him to be trivial and, second, that there is some genuine possibility that he will attain these goals. These observations are confirmed by various systematic studies of people who contemplated suicide, of others who unsuccessfully attempted suicide, and of situations in which people did commit suicide. When the subjects of these studies declared that their lives were no longer worth living they generally meant either that there was nothing left in their lives about which they seriously cared or that there was no real likelihood of attaining any of the goals that mattered to them. It should be noted that in this sense an individual may well be mistaken in his assertion that his life is or is not worthwhile any longer: he may, for example, mistake a temporary indisposition for a more permanent loss of interest, or, more likely, he may falsely estimate his chances of achieving the ends he wishes to attain.

DIFFERENT SENSES OF "WORTHWHILE"

According to the account given so far, one is saying much the same thing in declaring a life to be worthwhile and in asserting that it has meaning in the "terrestrial" sense of the word. There is, however, an interesting difference. When we say that a person's life has meaning (in the terrestrial sense) we are not committed to the claim that the goal or goals to which he is devoted have any positive value. (This is a slight oversimplification, assuming greater uniformity in the use of "meaning of life" than actually exists, but it will not seriously affect any of the controversial issues discussed here.) The question "As long as his life was dedicated to the spread of communism it had meaning *to him,* but was it really meaningful?" seems to be senseless. We are inclined to say, "If his life had meaning to him, then it had meaning—that's all there is to it." We are not inclined (or we are much less inclined) to say something of this kind when we speak of the worth of a person's life. We might say—for example, of someone like Eichmann—"While he was carrying out the extermination program, his life *seemed* worthwhile to him, but since his goal was so horrible, his life *was not* worthwhile." One might perhaps distinguish between a "subjective" and an "objective" sense of "worthwhile." In the subjective sense, saying that a person's life is worthwhile simply means that he is attached to some goals which he does not consider trivial and that these goals are attainable for him. In declaring that somebody's life is worthwhile in the objective sense, one is saying that he is attached to certain goals which are both attainable and of positive value.

It may be held that unless one accepts some kind of rationalist or intuitionist view of fundamental value judgments one would have to conclude that in the objective sense of "worthwhile" no human life (and indeed no human action) could ever be shown to be worthwhile. There is no need to enter here into a discussion of any controversial questions about the logical status of fundamental value judgments. But it may be pointed out that somebody who favors a subjectivist or emotivist account can quite consistently allow for the distinction between ends that only seem to have positive value and those that really do. To mention just one way in which this could be done: one may distinguish between ends that would be approved by rational and sympathetic human beings and those that do not carry such an endorsement. One may then argue that when we condemn such a life as Eichmann's as not being worthwhile we mean not that the ends to which he devoted himself possess some non-natural characteristic of badness but that no rational or sympathetic person would approve of them.

THE PESSIMISTS' SPECIAL STANDARDS

The unexciting conclusion of this discussion is that some human lives are at certain times not worthwhile in either of the two senses we have distinguished, that some are worthwhile in the subjective but not in the objective sense, some in the objective but not in the subjective sense, and some are worthwhile in both senses. The unexcitingness of this conclusion is not a reason for rejecting it, but some readers may question whether it meets the challenge of the pessimists. The pessimist, it may be countered, surely does not deny the plain fact that human beings are on occasions attached to goals which do not seem to them trivial, and it is also not essential to his position to deny (and most pessimists do not in fact deny) that these goals are sometimes attainable. The pessimist may even allow that in a superficial ("immediate") sense the goals which people try to achieve are of positive value, but he would add that because our lives are not followed by eternal bliss they are not "really" or "ultimately" worthwhile. If this is so, then the situation may be characterized by saying that the ordinary man and the pessimist do not mean the same by "worthwhile," or that they do mean the same in that both use it as a positive value expression but that their standards are different: the standards of the pessimist are very much more demanding than those of most ordinary people.

Anybody who agrees that death is final will have to concede that the pessimist is not mistaken in his contention that judged by *his* standards, life is never worthwhile. However, the pessimist is mistaken if he concludes, as frequently happens, that life is not worthwhile by ordinary standards because it is not worthwhile by his standards. Furthermore, setting aside the objection mentioned earlier (that there is something arbitrary about maintaining that eternal bliss makes life worthwhile but not allowing this role to bliss in the present life), one may justifiably ask why one should abandon ordinary standards in favor of those of the pessimist. Ordinarily, when somebody changes standards (for example, when a school raises or lowers its standards of admission) such a change can be supported by reasons. But how can the pessimist justify his special standards? It should be pointed out here that our ordinary standards do something for us which the pessimist's standards do not: they guide our choices, and as long as we live we can hardly help making choices. It is true that in one type of situation the pessimist's standards also afford guidance— namely, in deciding whether to go on living. It is notorious, however, that whether or not they are, by their own standards, rational in this, most pessimists do not commit suicide. They are then faced with much the same choices as other people. In these situations their own demanding standards are of no use, and in fact they avail themselves of the ordinary standards. Schopenhauer, for example, believed that if he had hidden his antireligious views he would have had no difficulty in obtaining an academic appointment and other worldly honors. He may have been mistaken in this belief, but in any event his actions indicate that he regarded intellectual honesty as worthwhile in a sense in which worldly honors were not. Again, when Darrow had the choice between continuing as counsel for the Chicago and North Western Railway and taking on the defense of Eugene V. Debs and his harassed and persecuted American Railway Union, he did not hesitate to choose the latter, apparently regarding it as worthwhile to go to the assistance of the suppressed and not worthwhile to aid the suppressor. In other words, although no human action is worthwhile, some human actions and presumably some human lives are less unworthwhile than others.

IS THE UNIVERSE BETTER WITH HUMAN LIFE THAN WITHOUT IT?

We have not—at least not explicitly—discussed the claims of Schopenhauer, Eduard von Hartmann, and other pessimists that the nonexistence of the world would be better than its existence, by which they mean that a world without human life would be better than one with it.

ARGUMENTS OF A PHENOMENOLOGIST

Some writers do not think that life can be shown to have meaning in any philosophically significant sense unless an affirmative answer to this question can be justified. Thus, in his booklet *Der Sinn unseres Daseins* the German phenomenologist Hans Reiner distinguishes between the everyday question about what he calls the "need-conditioned" meaning of life, which arises only for a person who is already in existence and has certain needs and desires, and the question about the meaning of human life in general. The latter question arises in concrete form when a responsible person is faced with the *Zeugungsproblem*—the question whether he should bring a child into the world. Reiner allows that a person's life has meaning in the former or "merely subjective" sense as long as his ordinary goals (chiefly his desire for happiness) are attained. This, however, does not mean that his life has an "objective" or "existential" (*seinshaft*) meaning—a significance or meaning which "attaches to life as such" and which, unlike the need-conditioned meaning, cannot be destroyed by any accident of fate. The philosopher, according to Reiner, is primarily concerned with the question of whether life has meaning in this objective or existential sense. "Our search for the meaning of our life," Reiner writes, "is identical with the search for a logically compelling reason (*einen einsichtigen Grund*) why it is better for us to exist than not to exist" (*Der Sinn unseres Daseins,* p. 27). Again, the real question is "whether it is better that mankind should exist than that there should be a world without any human life" (ibid., p. 31). It may be questioned whether this is what anybody normally means when he asks whether life has any meaning, but Reiner certainly addresses himself to one of the questions raised by Schopenhauer and other pessimists that ought to be discussed here.

Reiner believes that he can provide a "logically compelling reason" why a world with human life is better than one without it. He begins by pointing out that men differ from animals by being, among other things, "moral individuals." To be a moral individual is to be part of the human community and to be actively concerned in the life of other human beings. It is indeed undeniable that people frequently fail to bring about the ends of morally inspired acts or wishes, but phenomenological analysis discloses that "the real moral value and meaning" of an act does not depend on the attainment of the "external goal." As Kant correctly pointed out, the decisive factor is "the good will," the moral intent or attitude. It is here that we find the existential meaning of life: "Since that which is morally good contains its meaning and value within itself, it follows that it is intrinsically worthwhile. The existence of what is morally good is therefore better than its nonexistence" (ibid., pp. 54–55). But the existence of what is morally good is essentially connected with the existence of free moral individuals, and hence it follows that the existence of human beings as moral agents is better than their nonexistence.

Unlike happiness, which constitutes the meaning of life in the everyday or need-conditioned sense, the morally good does not depend on the accidents of life. It is not within a person's power to be happy, but it is "essentially" (*grundsätzlich*) in everybody's power to do what is good. Furthermore, while all happiness is subjective and transitory, leaving behind it no more than a "melancholy echo," the good has eternal value. Nobody would dream of honoring and respecting a person for his happiness or prosperity. On the other hand, we honor every good deed and the expression of every moral attitude, even if it took place in a distant land and among a foreign people. If we discover a good act or a good attitude in an enemy we nevertheless respect it and cannot help deriving a certain satisfaction from its existence. The same is true of good deeds carried out in ages long past. In all this the essentially timeless nature of morality becomes evident. Good deeds cease to exist as historical events only; their value, on the other hand, has eternal reality and is collected as an indestructible "fund." This may be a metaphysical statement, but it is not a piece of "metaphysical speculation." It simply makes explicit what the experience of the morally good discloses to phenomenological analysis (ibid., pp. 55–57).

REPLIES TO REINER

There is a great deal in this presentation with which one could take issue. If one is not misled by the image of the ever-growing, indestructible "fund," one may wonder, for example, what could be meant by claiming that the value of a good deed is "eternal," other than that most human beings tend to approve of such an action regardless of when or where it took place. However, we are here concerned primarily with the question whether Reiner has met the challenge of the pessimists, and it seems clear that he has not. A pessimist like Schopenhauer or Darrow might provisionally grant the correctness of Reiner's phenomenological analysis of morality but still offer the following rejoinder: The inevitable misery of all or nearly all human beings is so great that even if in the course of their lives they have a chance to preserve their inner moral natures or their good will, the continued torture to which their lives condemn them would not be justified. Given the pessimist's estimate of human life, this is surely not an unreasonable rejoinder. Even without relying on the pessimist's description of human life, somebody while accepting Reiner's phenomenological analysis might reach the opposite conclusion. He might, for example, share the quietist strain of Schopenhauer's teachings and object to the whole hustle and bustle of life, concluding that the "peace of the all-sufficient nothing"—or, more literally, a universe without human life—was better in spite of the fact that moral deeds could not then be performed. Since he admits the "facts" of morality on which Reiner bases his case but considers the peace of the all-sufficient nothing more valuable than morality, it is not easy to see how an appeal to the latter would show him to be mistaken. What phenomenological analysis has not disclosed, to Reiner or, as far as is known, to anybody else, is that doing good is the only or necessarily the greatest value.

WHY THE PESSIMIST CANNOT BE ANSWERED

The conclusion suggests itself that the pessimist cannot here be refuted, not because what he says is true or even because we do not know who is right and who is wrong but because the question whether a universe with human life is better than one without it does not have any clear meaning unless it is interpreted as a request for a statement of personal preference. The situation seems to be somewhat similar to what we found in the case of the question "Is my life better than my death?" when asked in certain circumstances. In some contexts indeed when we talk about human life in general, the word "better" has a reasonably clear meaning. Thus, if it is maintained that life for the human race will be better than it is now after cancer and mental illness have been conquered, or that human life will be better (or worse) after religion has disappeared, we understand fairly well what is meant, what facts would decide the issue either way. However, we do not really know what would count as evidence for or against the statement "The existence of human life as such is better than its nonexistence." Sometimes it is claimed that the question has a fairly clear meaning, namely, whether happiness outweighs unhappiness. Thus, von Hartmann supports his answer that the nonexistence of human life is better than its existence, that in fact an inanimate world would be better than one with life, with the argument that as we descend the scale of civilization and "sensitivity," we reach ever lower levels of misery. "The individuals of the lower and poorer classes and of ruder nations," he writes, "are happier than those of the elevated and wealthier classes and of civilized nations, not indeed because they are poorer and have to endure more want and privations, but because they are coarser and duller" (*Philosophy of the Unconscious*, Vol. III, p. 76). The "brutes," similarly, are "happier (i.e., less miserable)" than man, because "the excess of pain which an animal has to bear is less than that which a man has to bear." The same principle holds within the world of animals and plants:

> How much more painful is the life of the more finely-feeling horse compared with that of the obtuse pig, or with that of the proverbially happy fish in the water, its nervous system being of a grade so far inferior! As the life of a fish is more enviable than that of a horse, so is the life of an oyster than that of a fish, and the life of a plant than that of an oyster. (ibid.)

The conclusion is inevitable: the best or least undesirable form of existence is reached when, finally, we "descend beneath the threshold of consciousness"; for only there do we "see individual pain entirely disappear" (*Philosophy of the Unconscious*, Vol. III, pp. 76–77). Schopenhauer, also, addressing himself directly to the *"Zeugungsproblem,"* reaches a negative answer on the ground that unhappiness usually or necessarily outweighs happiness. "Could the human race continue to exist," he asks (in *Parerga und Paralipomena*, Vol. II, pp. 321–322), if "the generative act were . . . an affair of pure rational reflection? Would not rather everyone have so much compassion for the coming generation as to prefer to spare it the burden of existence, or at least be unwilling to take on himself the responsibility of imposing such a burden in cold blood?" In these passages Schopenhauer and von Hartmann assume that in the question "Is a world with human life better than one without human life?" the word "better" must be construed in a hedonistic or utilitarian sense—and the same is true of several other philosophers who do not adopt their pessimistic answer. However, while one may *stipulate* such a sense for "better" in this context, it is clear that this is *not* what is meant prior to the stipulation. Spinoza, for example, taught that the most miserable form of existence is preferable to nonexistence. Perhaps few who have directly observed the worst agonies and tortures that may be the lot of human beings or of animals would subscribe to this judgment, but Spinoza can hardly be accused of a self-contradictory error. Again, Nietzsche's philosophy is usually and quite accurately described as an affirmation of life, but Nietzsche was very careful not to play down the horrors of much of life. While he did not endorse Schopenhauer's value judgments, he thought that, by and large, Schopenhauer had not been far wrong in his description of the miseries of the human scene. In effect Nietzsche maintained that even though unhappiness is more prevalent than happiness, the existence of life is nevertheless better than its nonexistence, and this surely is not a self-contradiction.

It is important to point out what does not follow from the admission that in a nonarbitrary sense of "better," the existence of the human race cannot be shown to be better than its nonexistence: It does not follow that I or anybody else cannot or should not prefer the continued existence of the human race to its nonexistence or my own life to my death, and it does not follow that I or anybody else cannot or should not enjoy himself or that I or anybody else is "irrational" in any of these preferences. It is also impossible to prove that in some nonarbitrary sense of "better," coffee with cream is better than black coffee, but it does not follow that I cannot or should not prefer or enjoy it or that I am irrational in doing so. There is perhaps something a trifle absurd and obsessive in the need for a "proof" that the existence of life is better than its nonexistence. It resembles the demand to have it "established by argument" that love is better than hate.

Perhaps it would be helpful to summarize the main conclusions reached in this essay:

(1) In certain familiar senses of "meaning," which are not usually regarded as trivial, an action or a human life can have meaning quite independently of whether there is a God or whether we shall live forever.

(2) Writers like Tolstoy, who, because of the horror that death inspires, conclude that death is better than life, are plainly inconsistent. Moreover, the whole question of whether my life is better than my death, unless it is a question about my preference, seems to be devoid of sense.

(3) Those who argue that no human action can be worthwhile because we all must eventually die ignore what may be called the "short-term context" of much of our lives.

(4) Some human lives are worthwhile in one or both of the two senses in which "worthwhile" is commonly used, when people raise the question of whether a given person's life is worthwhile. The pessimists who judge human life by more demanding standards are not mistaken when they deny that by *their* standards no human life is ever worthwhile. However, they are guilty of a fallacious inference if they conclude that for this reason no human life can be worthwhile by the usual standards. Nor is it clear why anybody should embrace their standards in the place of those commonly adopted.

(5) It appears that the pessimists cannot be answered if in order to answer them one has to be able to prove that in some nonarbitrary sense of the word "better," the existence of life is better than its nonexistence. But this admission does not have any of the gloomy consequences which it is sometimes believed to entail.

RICHARD TAYLOR

The Meaning of Life

The question whether life has any meaning is difficult to interpret, and the more one concentrates his critical faculty on it the more it seems to elude him, or to evaporate as any intelligible question. One wants to turn it aside, as a source of embarrassment, as something that, if it cannot be abolished, should at least be decently covered. And yet I think any reflective person recognizes that the question it raises is important, and that it ought to have a significant answer.

If the idea of meaningfulness is difficult to grasp in this context, so that we are unsure what sort of thing would amount to answering the question, the idea of meaninglessness is perhaps less so. If, then, we can bring before our minds a clear image of meaningless existence, then perhaps we can take a step toward coping with our original question by seeing to what extent our lives, as we actually find them, resemble that image, and draw such lessons as we are able to from the comparison.

MEANINGLESS EXISTENCE

A perfect image of meaninglessness, of the kind we are seeking, is found in the ancient myth of Sisyphus. Sisyphus, it will be remembered, betrayed divine secrets to mortals, and for this he was condemned by the gods to roll a stone to the top of a hill, the stone then immediately to roll back down, again to be pushed to the top by Sisyphus, to roll down once more, and so on again and again, *forever*. Now in this we have the picture of meaningless, pointless toil, of a meaningless existence that is absolutely *never* redeemed. It is not even redeemed by a death that, if it were to accomplish nothing more, would at least bring this idiotic cycle to a close. If we were invited to imagine Sisyphus struggling for awhile and accomplishing nothing, perhaps eventually falling from exhaustion, so that we might suppose him then eventually turning to something having some sort of promise, then the meaninglessness of that chapter of his life would not be so stark. It would be a dark and dreadful dream, from which he eventually awakens to sunlight and reality. But he does not awaken, for there is nothing for him to awaken to. His repetitive toil is his life and reality, and it goes on forever, and it is without any meaning whatever. Nothing ever comes of what he is doing, except simply, more of the same. Not by one step, nor by a thousand, nor by ten thousand does he even expiate by the smallest token the sin against the gods that led him into this fate. Nothing comes of it, nothing at all.

This ancient myth has always enchanted men, for countless meanings can be read into it. Some of the ancients apparently thought it symbolized the perpetual rising and setting of the sun, and others the repetitious crashing of the waves upon the shore. Probably the commonest interpretation is that it symbolizes man's eternal struggle and unquenchable spirit, his determination always to try once more in the face of overwhelming discouragement. This interpretation is further supported by that version of the myth according to which Sisyphus was commanded to roll the stone *over* the hill, so that it would finally roll down the other side, but was never quite able to make it.

I am not concerned with rendering or defending any interpretation of this myth, however. I have cited it only for the one element it does unmistakably contain, namely, that of a repetitious, cyclic activity that never comes to anything. We could contrive other images of this that would serve just as well, and no myth-makers are needed to supply the materials of it. Thus, we can imagine two persons transporting a stone—or even a precious gem, it does not matter—back and forth, relay style. One carries it to a near or distant point where it is received by the other; it is returned to its starting point, there to be recovered by the first, and the process is repeated over and over. Except in this relay nothing counts as winning, and nothing brings the contest to any close, each step only leads to a repetition of itself. Or we can imagine two groups of prisoners, one of them engaged in digging a prodigious hole in the ground that is no sooner finished than it is filled in again by the other group, the latter then digging a new hole that is at once filled in by the first group, and so on and on endlessly.

Now what stands out in all such pictures as oppressive and dejecting is not that the beings who enact these roles suffer any torture or pain, for it need not be assumed that they do. Nor is it that their labors are great, for they are no greater than the labors commonly undertaken by most men most of the time. According to the original myth, the stone is so large that Sisyphus never quite gets it to the top and must groan under every step, so that his enormous labor is all for nought. But this is not what appalls. It is not that his great struggle comes to nothing, but that his existence itself is without meaning. Even if we suppose, for example, that the stone is but a pebble that can be carried effortlessly, or that the holes dug by the prisoners are but small ones, not the slightest meaning is introduced into their lives. The stone that Sisyphus moves to the top of the hill, whether we think of it as large or small, still rolls back every time, and the process is repeated forever. Nothing comes of it, and the work is simply pointless. That is the element of the myth that I wish to capture.

Again, it is not the fact that the labors of Sisyphus continue forever that deprives them of meaning. It is, rather, the implication of this: that they come to nothing. The image would not be changed by our supposing him to push a different stone up every time, each to roll down again. But if we supposed that these stones, instead of rolling back to their places as if they had never been moved, were assembled at the top of the hill and there incorporated, say, in a beautiful and enduring temple, then the aspect of meaninglessness would disappear. His labors would then have a point, something would come of them all, and although one could perhaps still say it was not worth it, one could not say that the life of Sisyphus was devoid of meaning altogether. Meaningfulness would at least have made an appearance, and we could see what it was.

That point will need remembering. But in the meantime, let us note another way in which the image of meaninglessness can be altered by making only a very slight change. Let us suppose that the gods, while condemning Sisyphus to the fate just described, at the same time, as an afterthought, waxed perversely merciful by implanting in him a strange and irrational impulse; namely, a compulsive impulse to roll stones. We may if we like, to make this more graphic, suppose they accomplish this by implanting in him some substance that has this effect on his character and drives. I call this perverse, because from our point of view there is clearly no reason why anyone should have a persistent and insatiable desire to do something so pointless as that. Nevertheless, suppose that is Sisyphus' condition. He has but one obsession, which is to roll stones, and it is an obsession that is only for the moment appeased by his rolling them—he no sooner gets a stone rolled to the top of the hill than he is restless to roll up another.

Now it can be seen why this little afterthought of the gods, which I called perverse, was also in fact merciful. For they have by this device managed to give Sisyphus precisely what he wants—by making him want precisely what they inflict on him. However it may appear to us, Sisyphus' fate now does not appear to him as a condemnation, but the very reverse. His one desire in life is to roll stones, and he is absolutely guaranteed its endless fulfillment. Where otherwise he might profoundly have wished surcease, and even welcomed the quiet of death to release him from endless boredom and meaninglessness, his life is now filled with mission and meaning, and he seems to himself to have been given an entry to heaven. Nor need he even fear death, for the gods have promised him an endless opportunity to indulge his single purpose, without concern or frustration. He will be able to roll stones *forever*.

What we need to mark most carefully at this point is that the picture with which we began has not really been changed in the least by adding this supposition. Exactly the same things happen as before. The only change is in Sisyphus' view of them. The picture before was the image of meaningless activity and existence. It was created precisely to be an image of that. It has not lost that meaninglessness, it has now gained not the least shred of meaningfulness. The stones still roll back as before, each phase of Sisyphus' life still exactly resembles all the others, the task is never completed, nothing comes of it, no temple ever begins to rise, and all this cycle of the same pointless thing over and over goes on forever in this picture as in the other. The *only* thing that has happened is this: Sisyphus has been reconciled to it, and indeed more, he has been led to embrace it. Not, however, by reason or persuasion, but by nothing more rational than the potency of a new substance in his veins.

THE MEANINGLESSNESS OF LIFE

I believe the foregoing provides a fairly clear content to the idea of meaninglessness and, through it, some hint of what meaningfulness, in this sense, might be. Meaninglessness is essentially endless pointlessness, and meaningfulness is therefore the opposite. Activity, and even long, drawn-out and repetitive activity, has a meaning if it has some significant culmination, some more or less lasting end that can be considered to have been the direction and purpose of the activity. But the descriptions so far also provide something else; namely, the suggestion of how an existence that is objectively meaningless, in this sense, can nevertheless acquire a meaning for him whose existence it is.

Now let us ask: Which of these pictures does life in fact resemble? And let us not begin with our own lives, for here both our prejudices and wishes are great, but with the life in general that we share with the rest of creation. We shall find, I think, that it all has a certain pattern, and that this pattern is by now easily recognized.

We can begin anywhere, only saving human existence for our last consideration. We can, for example, begin with any animal. It does not matter where we begin, because the result is going to be exactly the same.

Thus, for example, there are caves in New Zealand, deep and dark, whose floors are quiet pools and whose walls and ceilings are covered with soft light. As one gazes in wonder in the stillness of these caves it seems that the Creator has reproduced there in microcosm the heavens themselves, until one scarcely remembers the enclosing presence of the walls. As one looks more closely, however, the scene is explained. Each dot of light identifies an ugly worm, whose luminous tail is meant to attract insects from the surrounding darkness. As from time to time one of these insects draws near it becomes entangled in a sticky thread lowered by the worm, and is eaten. This goes on month after month, the blind worm lying there in the barren stillness waiting to entrap an occasional bit of nourishment that will only sustain it to another bit of nourishment until. . . . Until what? What great thing awaits all this long and repetitious effort and makes it worthwhile? Really nothing. The larva just transforms itself finally to a tiny winged adult that lacks even mouth parts to feed and lives only a day or two. These adults, as soon as they have mated and laid eggs, are themselves caught in the threads and are devoured by the cannibalist worms, often without having ventured into the day, the only point to their existence having now been fulfilled. This has been going on for millions of years, and to no end other than that the same meaningless cycle may continue for another millions of years.

All living things present essentially the same spectacle. The larva of a certain cicada burrows in the darkness of the earth for seventeen years, through season after season, to emerge finally into the daylight for a brief flight, lay its eggs, and die—this all to repeat itself during the next seventeen years, and so on to eternity. We have already noted, in another connection, the struggles of fish, made only that others may do the same after them and that this cycle, having no other point than itself, may never cease. Some birds span an entire side of the globe each year and then return, only to insure that others may follow the same incredibly long path again and again. One is led to wonder what the point of it all is, with what great triumph this ceaseless effort, repeating itself through millions of years, might finally culminate, and why it should go on and on for so long, accomplishing nothing, getting nowhere. But then one realizes that there is no point to it at all, that it really culminates in nothing, that each of these cycles, so filled with toil, is to be followed only by more of the same. The point of any living thing's life is, evidently, nothing but life itself.

This life of the world thus presents itself to our eyes as a vast machine, feeding on itself, running on and on forever to nothing. And we are part of that life. To be sure, we are not just the same, but the differences are not so great as we like to think; many are merely invented, and none really cancels the kind of meaninglessness that we found in Sisyphus and that we find all around, wherever anything lives. We are conscious of our activity. Our goals, whether in any significant sense we choose them or not, are things of which we are at least partly aware and can therefore in some sense appraise. More significantly, perhaps, men have a history, as other animals do not, such that each generation does not precisely resemble all those before. Still, if we can in imagination disengage our wills from our lives and disregard the deep interest each man has in his own existence, we shall find that they do not so little resemble the existence of Sisyphus. We toil after goals, most of them—indeed every single one of them—of transitory significance and, having gained one of them, we immediately set forth for the next, as if that one had never been, with this next one being essentially more of the same. Look at a busy street any day, and observe the throng going hither and thither. To what? Some office or shop, where the same things will be done today as were done yesterday, and are done now so they may be repeated tomorrow. And if we think that, unlike Sisyphus, these labors do have a point, that they culminate in something lasting and, independently of our own deep interests in them, very worthwhile, then we simply have not considered the thing closely enough. Most such effort is directed only to the establishment and perpetuation of home and family; that is, to the begetting of others who will follow in our steps to do more of the same. Each man's life thus resembles one of Sisyphus' climbs to the summit of his hill, and each day of it one of his steps; the difference is that whereas Sisyphus himself returns to push the stone up again, we leave this to our children. We at one point imagined that the labors of Sisyhus finally culminated in the creation of a temple, but for this to make any difference it had to be a temple that would at least endure, adding beauty to the world for the remainder of time. Our achievements, even though they are often beautiful, are mostly bubbles; and those that do last, like the sand-swept pyramids, soon becomes mere curiosities while around them the rest of mankind continues its perpetual toting of rocks, only to see them roll down. Nations are built upon the bones of their founders and pioneers, but only to decay and crumble before long, their rubble then becoming the foundation for others directed to exactly the same fate. The picture of Sisyphus is the picture of existence of the individual man, great or unknown, of nations, of the race of men, and of the very life of the world.

On a country road one sometimes comes upon the ruined hulks of a house and once extensive buildings, all in collapse and spread over with weeds. A curious eye can in imagination reconstruct from what is left a once warm and thriving life, filled with purpose. There was the hearth, where a family once talked, sang, and made plans; there were the rooms, where people loved, and babes were born to a rejoicing mother; there are the musty remains of a sofa, infested with bugs, once bought at a dear price to enhance an ever-growing comfort, beauty, and warmth. Every small piece of junk fills the mind with what once, not long ago, was utterly real, with children's voices, plans made, and enterprises embarked upon. That is how these stones of Sisyphus were rolled up, and that is how they became incorporated into a beautiful temple, and that temple is what now lies before you. Meanwhile other buildings, institutions, nations, and civilizations spring up all around, only to share the same fate before long. And if the question "What for?" is now asked, the answer is clear: so that just this may go on forever.

The two pictures—of Sisyphus and of our own lives, if we look at them from a distance—are in outline the same and convey to the mind the same image. It is not surprising, then, that men invent ways of denying it, their religions proclaiming a heaven that does not crumble, their hymnals and prayer books declaring a significance to life of which our eyes provide no hint whatever.[1] Even our philosophies portray some permanent and lasting good at which all may aim, from the changeless forms invented by Plato to the beatific vision of St. Thomas and the ideals of permanence contrived by the moderns. When these fail to convince, then earthly ideals such as universal justice and brotherhood are conjured up to take their places and give meaning to man's seemingly endless pilgrimage, some final state that will be ushered in when the last obstacle is removed and the last stone pushed to the hilltop. No one believes, of course, that any such state will be final, or even wants it to be in case it means that human existence would then cease to be a struggle; but in the meantime such ideas serve a very real need.

THE MEANING OF LIFE

We noted that Sisyphus' existence would have meaning if there were some point to his labors, if his efforts ever culminated in something that was not just an occasion for fresh labors of the same kind. But that is

---

1. A popular Christian hymn, sung often at funerals and typical of many hymns, expresses this thought:

> Swift to its close ebbs out life's little day;
> Earth's joys grow dim, its glories pass away;
> Change and decay in all around I see:
> O thou who changest not, abide with me.

precisely the meaning it lacks. And human existence resembles his in that respect. Men do achieve things—they scale their towers and raise their stones to their hilltops—but every such accomplishment fades, providing only an occasion for renewed labors of the same kind.

But here we need to note something else that has been mentioned, but its significance not explored, and that is the state of mind and feeling with which such labors are undertaken. We noted that if Sisyphus had a keen and unappeasable desire to be doing just what he found himself doing, then, although his life would in no way be changed, it would nevertheless have a meaning for him. It would be an irrational one, no doubt, because the desire itself would be only the product of the substance in his veins, and not any that reason could discover, but a meaning nevertheless.

And would it not, in fact, be a meaning incomparably better than the other? For let us examine again the first kind of meaning it could have. Let us suppose that, without having any interest in rolling stones, as such, and finding this, in fact, a galling toil, Sisyphus did nevertheless have a deep interest in raising a temple, one that would be beautiful and lasting. And let us suppose he succeeded in this, that after ages of dreadful toil, all directed at this final result, he did at last complete his temple, such that now he could say his work was done, and he could rest and forever enjoy the result. Now what? What picture now presents itself to our minds? It is precisely the picture of infinite boredom! Of Sisyphus doing nothing ever again, but contemplating what he has already wrought and can no longer add anything to, and contemplating it for an eternity! Now in this picture we have a meaning for Sisyphus' existence, a point for his prodigious labor, because we have put it there; yet, at the same time, that which is really worthwhile seems to have slipped away entirely. Where before we were presented with the nightmare of eternal and pointless activity, we are now confronted with the hell of its eternal absence.

Our second picture, then, wherein we imagined Sisyphus to have had inflicted on him the irrational desire to be doing just what he found himself doing, should not have been dismissed so abruptly. The meaning that picture lacked was no meaning that he or anyone could crave, and the strange meaning it had was perhaps just what we were seeking.

At this point, then, we can reintroduce what has been until now, it is hoped, resolutely pushed aside in an effort to view our lives and human existence with objectivity; namely, our own wills, our deep interest in what we find ourselves doing. If we do this we find that our lives do indeed still resemble that of Sisyphus, but that the meaningfulness they thus lack is precisely the meaningfulness of infinite boredom. At the same time, the strange meaningfulness they possess is that of the inner compulsion to be doing just what we were put here to do, and to go on doing it forever. This is the nearest we may hope to get to heaven, but the redeeming side of that fact is that we do thereby avoid a genuine hell.

If the builders of a great and flourishing ancient civilization could somehow return now to see archaeologists unearthing the trivial remnants of what they had once accomplished with such effort—see the fragments of pots and vases, a few broken statues, and such tokens of another age and greatness—they could indeed ask themselves what the point of it all was, if this is all it finally came to. Yet, it did not seem so to them then, for it was just the building, and not what was finally built, that gave their life meaning. Similarly, if the builders of the ruined home and farm that I described a short while ago could be brought back to see what is left, they would have the same feelings. What we construct in our imaginations as we look over these decayed and rusting pieces would reconstruct itself in their very memories, and certainly with unspeakable sadness. The piece of a sled at our feet would revive in them a warm Christmas. And what rich memories would there be in the broken crib? And the weed-covered remains of a fence would reproduce the scene of a great herd of livestock, so laboriously built up over so many years. What was it all worth, if this is the final result? Yet, again, it did not seem so to them through those many years of struggle and toil, and they did not imagine they were building a Gibraltar. The things to which they bent their backs day after day, realizing one by one their ephemeral plans, were precisely the things in which their wills were deeply involved, precisely the things in which their interests lay, and there was no need then to ask questions. There is no more need of them now—the day was sufficient to itself, and so was the life.

This is surely the way to look at all of life—at one's own life, and each day and moment it contains; of the life of a nation; of the species; of the life of the world; and of everything that breathes. Even the glow worms I described, whose cycles of existence over the millions of years seem so pointless when looked at by us, will seem entirely different to us if we can somehow try to view their existence from within. Their endless activity, which gets nowhere, is just what it is their will to pursue. This is its whole justification and meaning. Nor would it be any salvation to the birds who span the globe every year, back and forth, to have a home made for them in a cage with plenty of food and protection, so that they would not have to migrate any more. It would be their condemnation, for it is the doing that counts for them, and not what they hope to win by it. Flying these prodigious distances, never ending, is what it is in their veins to do, exactly as it was in Sisyphus' veins to roll stones, without end, after the gods had waxed merciful and implanted this in him.

A human being no sooner draws his first breath than he responds to the will that is in him to live. He no more asks whether it will be worthwhile, or whether anything of significance will come of it, than the worms and the birds. The point of his living is simply to be living, in the manner that it is his nature to be living. He goes through his life building his castles, each of these beginning to fade into time as the next is begun; yet, it would be no salvation to rest from all this. It would be a condemnation, and one that would in no way be redeemed were he able to gaze upon the things he has done, even if these were beautiful and absolutely permanent, as they never are. What counts is that one should be able to begin a new task, a new castle, a new bubble. It counts only because it is there to be done and he has the will to do it. The same will be the life of his children, and of theirs; and if the philosopher is apt to see in this a pattern similar to the unending cycles of the existence of Sisyphus, and to despair, then it is indeed because the meaning and point he is seeking is not there—but mercifully so. The meaning of life is from within us, it is not bestowed from without, and it far exceeds in both its beauty and permanence any heaven of which men have ever dreamed or yearned for.

THOMAS NAGEL

The Absurd

Most people feel on occasion that life is absurd, and some feel it vividly and continually. Yet the reasons usually offered in defense of this conviction are patently inadequate: they *could* not really explain why life is absurd. Why then do they provide a natural expression for the sense that it is?

I

Consider some examples. It is often remarked that nothing we do now will matter in a million years. But if that is true, then by the same token, nothing that will be the case in a million years matters now. In particular, it does not matter now that in a million years nothing we do now will matter. Moreover, even if what we did now *were* going to matter in a million years, how could that keep our present concerns from being absurd? If their mattering now is not enough to accomplish that, how would it help if they mattered a million years from now?

Whether what we do now will matter in a million years could make the crucial difference only if its mattering in a million years depended on its mattering, period. But then to deny that whatever happens now will matter in a million years is to beg the question against its mattering, period; for in that sense one cannot know that it will not matter in a million years whether (for example) someone now is happy or miserable, without knowing that it does not matter, period.

What we say to convey the absurdity of our lives often has to do with space or time: we are tiny specks in the infinite vastness of the universe; our lives are mere instants even on a geological time scale, let alone a cosmic one; we will all be dead any minute. But of course none of these evident facts can be what *makes* life absurd, if it is absurd. For suppose we lived forever; would not a life that is absurd if it lasts seventy years be infinitely absurd if it lasted through eternity? And if our lives are absurd given our present size, why would they be any less absurd if we filled the universe (either because we were larger or because the universe was smaller)? Reflection on our minuteness and brevity appears to be intimately connected with the sense that life is meaningless; but it is not clear what the connection is.

Another inadequate argument is that because we are going to die, all chains of justification must leave off in mid-air: one studies and works to earn money to pay for clothing, housing, entertainment, food, to sustain oneself from year to year, perhaps to support a family and pursue a career—but to what final end? All of it is an elaborate journey leading nowhere. (One will also have some effect on other people's lives, but that simply reproduces the problem, for they will die too.)

There are several replies to this argument. First, life does not consist of a sequence of activities each of which has as its purpose some later member of the sequence. Chains of justification come repeatedly to an end within life, and whether the process as a whole can be justified has no bearing on the finality of these end-points. No further justification is needed to make it reasonable to take aspirin for a headache, attend an exhibit of the work of a painter one admires, or stop a child from putting his hand on a hot stove. No larger context or further purpose is needed to prevent these acts from being pointless.

Even if someone wished to supply a further justification for pursuing all the things in life that are commonly regarded as self-justifying, that justification would have to end somewhere too. If *nothing* can justify unless it is justified in terms of something outside itself, which is also justified, then an infinite regress results, and no chain of justification can be complete. Moreover, if a finite chain of reasons cannot justify anything, what could be accomplished by an infinite chain, each link of which must be justified by something outside itself?

Since justifications must come to an end somewhere, nothing is gained by denying that they end where they appear to, within life—or by trying to subsume the multiple, often trivial ordinary justifications of action under a single, controlling life scheme. We can be satisfied more easily than that. In fact, through its misrepresentation of the process of justification, the argument makes a vacuous demand. It insists that the reasons available within life are incomplete, but suggests thereby that all reasons that come to an end are incomplete. This makes it impossible to supply any reasons at all.

The standard arguments for absurdity appear therefore to fail as arguments. Yet I believe they attempt to express something that is difficult to state, but fundamentally correct.

II

In ordinary life a situation is absurd when it includes a conspicuous discrepancy between pretension or aspiration and reality: someone gives a complicated speech in support of a motion that has already been passed; a notorious criminal is made president of a major philanthropic foundation; you declare your lover over the telephone to a recorded announcement; as you are being knighted, your pants fall down.

When a person finds himself in an absurd situation, he will usually attempt to change it, by modifying his aspirations, or by trying to bring reality into better accord with them, or by removing himself from the situation entirely. We are not always willing or able to extricate ourselves from a position whose absurdity has become clear to us. Nevertheless, it is usually possible to imagine some change that would remove the absurdity—whether or not we can or will implement it. The sense that life as a whole is absurd arises when we perceive, perhaps dimly, an inflated pretension or aspiration which is inseparable from the continuation of human life and which makes its absurdity inescapable, short of escape from life itself.

Many people's lives are absurd, temporarily or permanently, for conventional reasons having to do with their particular ambitions, circumstances, and personal relations. If there is a philosophical sense of absurdity, however, it must arise from the perception of something universal—some respect in which pretension and reality inevitably clash for us all. This condition is supplied, I shall argue, by the collision between the seriousness with which we take our lives and the perpetual possibility of regarding everything about which we are serious as arbitrary, or open to doubt.

We cannot live human lives without energy and attention, nor without making choices which show that we take some things more seriously than others. Yet we have always available a point of view outside the particular form of our lives, from which the seriousness appears gratuitous. These two inescapable viewpoints collide in us, and that is what makes life absurd. It is absurd because we ignore the doubts that we know cannot be settled, continuing to live with nearly undiminished seriousness in spite of them.

This analysis requires defense in two respects: first as regards the unavoidability of seriousness; second as regards the inescapability of doubt.

We take ourselves seriously whether we lead serious lives or not and whether we are concerned primarily with fame, pleasure, virtue, luxury, triumph, beauty, justice, knowledge, salvation, or mere survival. If we take other people seriously and devote ourselves to them, that only multiplies the problem. Human life is full of effort, plans, calculation, success and failure: we *pursue* our lives, with varying degrees of sloth and energy.

It would be different if we could not step back and reflect on the process, but were merely led from impulse to impulse without self-consciousness. But human beings do not act solely on impulse. They are prudent, they reflect, they weigh consequences, they ask whether what they are doing is worth while. Not only are their lives full of particular choices that hang together in larger activities with temporal structure: they also decide in the broadest terms what to pursue and what to avoid, what the priorities among their various aims should be, and what kind of people they want to be or become. Some men are faced with such choices by the large decisions they make from time to time; some merely by reflection on the course their lives are taking as the product of countless small decisions. They decide whom to marry, what profession to follow, whether to join the Country Club, or the Resistance; or they may just wonder why they go on being salesmen or academics or taxi drivers, and then stop thinking about it after a certain period of inconclusive reflection.

Although they may be motivated from act to act by those immediate needs with which life presents them, they allow the process to continue by adhering to the general system of habits and the form of life in which such motives have their place—or perhaps only by clinging to life itself. They spend enormous quantities of energy, risk, and calculation on the details. Think of how an ordinary individual sweats over his appearance, his health, his sex life, his emotional honesty, his social utility, his self-knowledge, the quality of his ties with family, colleagues, and friends, how well he does his job, whether he understands the world and what is going on in it. Leading a human life is a full-time occupation, to which everyone devotes decades of intense concern.

This fact is so obvious that it is hard to find it extraordinary and important. Each of us lives his own life—lives with himself twenty-four hours a day. What else is he supposed to do—live someone else's life? Yet humans have the special capacity to step back and survey themselves, and the lives to which they are committed, with that detached amazement which comes from watching an ant struggle up a heap of sand. Without developing the illusion that they are able to escape from their highly specific and idiosyncratic position, they can view it *sub specie aeternitatis*—and the view is at once sobering and comical.

The crucial backward step is not taken by asking for still another justification in the chain, and failing to get it. The objections to that line of attack have already been stated; justifications come to an end. But this is precisely what provides universal doubt with its object. We step back to find that the whole system of justification and criticism, which controls our choices and supports our claims to rationality, rests on responses and habits that we never question, that we should not know how to defend without circularity, and to which we shall continue to adhere even after they are called into question.

The things we do or want without reasons, and without requiring reasons—the things that define what is a reason for us and what is not—are the starting points of our skepticism. We see ourselves from outside, and all the contingency and specificity of our aims and pursuits become clear. Yet when we take this view and recognize what we do as arbitrary, it does not disengage us from life, and there lies our absurdity: not in the fact that such an external view can be taken of us, but in the fact that we ourselves can take it, without ceasing to be the persons whose ultimate concerns are so coolly regarded.

III

One may try to escape the position by seeking broader ultimate concerns, from which it is impossible to step back—the idea being that absurdity results because what we take seriously is something small and insignificant and individual. Those seeking to supply their lives with meaning usually envision a role or function in something larger than themselves. They therefore seek fulfillment in service to society, the state, the revolution, the progress of history, the advance of science, or religion and the glory of God.

But a role in some larger enterprise cannot confer significance unless that enterprise is itself significant. And its significance must come back to what we can understand, or it will not even appear to give us what we are seeking. If we learned that we were being raised to provide food for other creatures fond of human flesh, who planned to turn us into cutlets before we got too stringy—even if we learned that the human race had been developed by animal breeders precisely for this purpose—that would still not give our lives meaning, for two reasons. First, we would still be in the dark as to the significance of the lives of those other beings; second, although we might acknowledge that this culinary role would make our lives meaningful to them, it is not clear how it would make them meaningful to us.

Admittedly, the usual form of service to a higher being is different from this. One is supposed to behold and partake of the glory of God, for example, in a way in which chickens do not share in the glory of coq au vin. The same is true of service to a state, a movement, or a revolution. People can come to feel, when they are part of something bigger, that it is part of them too. They worry less about what is peculiar to themselves, but identify enough with the larger enterprise to find their role in it fulfilling.

However, any such larger purpose can be put in doubt in the same way that the aims of an individual life can be, and for the same reasons. It is as legitimate to find ultimate justification there as to find it earlier, among the details of individual life. But this does not alter the fact that justifications come to an end when we are content to have them end—when we do not find it necessary to look any further. If we can step back from the purposes of individual life and doubt their point, we can step back also from the progress of human history, or of science, or the success of a society, or the kingdom, power, and glory of God,[1] and put all these things into question in the same way. What seems to us to confer meaning, justification, significance, does so in virtue of the fact that we need no more reasons after a certain point.

What makes doubt inescapable with regard to the limited aims of individual life also makes it inescapable with regard to any larger purpose that encourages the sense that life is meaningful. Once the fundamental doubt has begun, it cannot be laid to rest.

Camus maintains in *The Myth of Sisyphus* that the absurd arises because the world fails to meet our demands for meaning. This suggests that the world might satisfy those demands if it were different. But now we can see that this is not the case. There does not appear to be any conceivable world (containing us) about which unsettlable doubts could not arise. Consequently the absurdity of our situation derives not from a collision between our expectations and the world, but from a collision within ourselves.

IV

It may be objected that the standpoint from which these doubts are supposed to be felt does not exist—that if we take the recommended backward step we will land on thin air, without any basis for judgment about the natural responses we are supposed to be surveying. If we retain our usual standards of what is important, then questions about the significance of what we are doing with our lives will be answerable in the usual way. But if we do not, then those questions can mean nothing to us, since there

1. Cf. Robert Nozick, "Teleology," *Mosaic*, XII, 1 (Spring 1971): 27/8.

is no longer any content to the idea of what matters, and hence no content to the idea that nothing does.

But this objection misconceives the nature of the backward step. It is not supposed to give us an understanding of what is *really* important, so that we see by contrast that our lives are insignificant. We never, in the course of these reflections, abandon the ordinary standards that guide our lives. We merely observe them in operation, and recognize that if they are called into question we can justify them only by reference to themselves, uselessly. We adhere to them because of the way we are put together; what seems to us important or serious or valuable would not seem so if we were differently constituted.

In ordinary life, to be sure, we do not judge a situation absurd unless we have in mind some standards of seriousness, significance, or harmony with which the absurd can be contrasted. This contrast is not implied by the philosophical judgment of absurdity, and that might be thought to make the concept unsuitable for the expression of such judgments. This is not so, however, for the philosophical judgment depends on another contrast which makes it a natural extension from more ordinary cases. It departs from them only in contrasting the pretensions of life with a larger context in which *no* standards can be discovered, rather than with a context from which alternative, overriding standards may be applied.

V

In this respect, as in others, philosophical perception of the absurd resembles epistemological skepticism. In both cases the final, philosophical doubt is not contrasted with any unchallenged certainties, though it is arrived at by extrapolation from examples of doubt within the system of evidence or justification, where a contrast with other certainties *is* implied. In both cases our limitedness joins with a capacity to transcend those limitations in thought (thus seeing them as limitations, and as inescapable).

Skepticism begins when we include ourselves in the world about which we claim knowledge. We notice that certain types of evidence convince us, that we are content to allow justifications of belief to come to an end at certain points, that we feel we know many things even without knowing or having grounds for believing the denial of others which, if true, would make what we claim to know false.

For example, I know that I am looking at a piece of paper, although I have no adequate grounds to claim I know that I am not dreaming; and if I am dreaming then I am not looking at a piece of paper. Here an ordinary conception of how appearance may diverge from reality is em-

ployed to show that we take our world largely for granted; the certainty
that we are not dreaming cannot be justified except circularly, in terms of
those very appearances which are being put in doubt. It is somewhat far-
fetched to suggest I may be dreaming; but the possibility is only illustra-
tive. It reveals that our claim to knowledge depend on our not feeling it
necessary to exclude certain incompatible alternatives, and the dreaming
possibility or the total-hallucination possibility are just representatives for
limitless possibilities most of which we cannot even conceive.[2]

Once we have taken the backward step to an abstract view of our
whole system of beliefs, evidence, and justification, and seen that it works
only, despite its pretensions, by taking the world largely for granted, we
are *not* in a position to contrast all these appearances with an alternative
reality. We cannot shed our ordinary responses, and if we could it would
leave us with no means of conceiving a reality of any kind.

It is the same in the practical domain. We do not step outside our
lives to a new vantage point from which we see what is really, objectively
significant. We continue to take life largely for granted while seeing that
all our decisions and certainties are possible only because there is a great
deal we do not bother to rule out.

Both epistemological skepticism and a sense of the absurd can be
reached via initial doubts posed within systems of evidence and justification that we accept, and can be stated without violence to our ordinary
concepts. We can ask not only why we should believe there is a floor
under us, but also why we should believe the evidence of our senses at
all—and at some point the framable questions will have outlasted the
answers. Similarly, we can ask not only why we should take aspirin, but
why we should take trouble over our own comfort at all. The fact that we
shall take the aspirin without waiting for an answer to this last question
does not show that it is an unreal question. We shall also continue to believe there is a floor under us without waiting for an answer to the other
question. In both cases it is this unsupported natural confidence that generates skeptical doubts; so it cannot be used to settle them.

Philosophical skepticism does not cause us to abandon our ordinary
beliefs, but it lends them a peculiar flavor. After acknowledging that their
truth is incompatible with possibilities that we have no grounds for believing do not obtain—apart from grounds in those very beliefs which we
have called into question—we return to our familiar convictions with a
certain irony and resignation. Unable to abandon the natural responses on

2. I am aware that skepticism about the external world is widely thought to have
been refuted, but I have remained convinced of its irrefutability since being exposed
at Berkeley to Thompson Clarke's largely unpublished ideas on the subject.

which they depend, we take them back, like a spouse who has run off with someone else and then decided to return; but we regard them differently (not that the new attitude is necessarily inferior to the old, in either case).

The same situation obtains after we have put in question the seriousness with which we take our lives and human life in general and have looked at ourselves without presuppositions. We then return to our lives, as we must, but our seriousness is laced with irony. Not that irony enables us to escape the absurd. It is useless to mutter: "Life is meaningless; life is meaningless . . ." as an accompaniment to everything we do. In continuing to live and work and strive, we take ourselves seriously in action no matter what we say.

What sustains us, in belief as in action, is not reason or justification, but something more basic than these—for we go on in the same way even after we are convinced that the reasons have given out.[3] If we tried to rely entirely on reason, and pressed it hard, our lives and beliefs would collapse—a form of madness that may actually occur if the inertial force of taking the world and life for granted is somehow lost. If we lose our grip on that, reason will not give it back to us.

VI

In viewing ourselves from a perspective broader than we can occupy in the flesh, we become spectators of our own lives. We cannot do very much as pure spectators of our own lives, so we continue to lead them, and devote ourselves to what we are able at the same time to view as no more than a curiosity, like the ritual of an alien religion.

This explains why the sense of absurdity finds its natural expression in those bad arguments with which the discussion began. Reference to our small size and short lifespan and to the fact that all of mankind will eventually vanish without a trace are metaphors for the backward step which permits us to regard ourselves from without and to find the particular form of our lives curious and slightly surprising. By feigning a nebula's-eye

3. As Hume says in a famous passage of the *Treatise:* "Most fortunately it happens, that since reason is incapable of dispelling these clouds, nature herself suffices to that purpose, and cures me of this philosophical melancholy and delirium, either by relaxing this bent of mind, or by some avocation, and lively impression of my senses, which obliterate all these chimeras. I dine, I play a game of backgammon, I converse, and am merry with my friends; and when after three or four hours' amusement, I would return to these speculations, they appear so cold, and strain'd, and ridiculous, that I cannot find in my heart to enter into them any farther" (Book 1, Part 4, Section 7; Selby-Bigge, p. 269).

view, we illustrate the capacity to see ourselves without presuppositions, as arbitrary, idiosyncratic, highly specific occupants of the world, one of countless possible forms of life.

Before turning to the question whether the absurdity of our lives is something to be regretted and if possible escaped, let me consider what would have to be given up in order to avoid it.

Why is the life of a mouse not absurd? The orbit of the moon is not absurd either, but that involves no strivings or aims at all. A mouse, however, has to work to stay alive. Yet he is not absurd, because he lacks the capacities for self-consciousness and self-transcendence that would enable him to see that he is only a mouse. If that *did* happen, his life would become absurd, since self-awareness would not make him cease to be a mouse and would not enable him to rise above his mousely strivings. Bringing his new-found self-consciousness with him, he would have to return to his meagre yet frantic life, full of doubts that he was unable to answer, but also full of purposes that he was unable to abandon.

Given that the transcendental step is natural to us humans, can we avoid absurdity by refusing to take that step and remaining entirely within our sublunar lives? Well, we cannot refuse consciously, for to do that we would have to be aware of the viewpoint we were refusing to adopt. The only way to avoid the relevant self-consciousness would be either never to attain it or to forget it—neither of which can be achieved by the will.

On the other hand, it is possible to expend effort on an attempt to destroy the other component of the absurd—abandoning one's earthly, individual, human life in order to identify as completely as possible with that universal viewpoint from which human life seems arbitrary and trivial. (This appears to be the ideal of certain Oriental religions.) If one succeeds, then one will not have to drag the superior awareness through a strenuous mundane life, and absurdity will be diminished.

However, insofar as this self-etiolation is the result of effort, willpower, asceticism, and so forth, it requires that one take oneself seriously as an individual—that one be willing to take considerable trouble to avoid being creaturely and absurd. Thus one may undermine the aim of unworldliness by pursuing it too vigorously. Still, if someone simply allowed his individual, animal nature to drift and respond to impulse, without making the pursuit of its needs a central conscious aim, then he might, at considerable dissociative cost, achieve a life that was less absurd than most. It would not be a meaningful life either, of course; but it would not involve the engagement of a transcendent awareness in the assiduous pursuit of mundane goals. And that is the main condition of absurdity—the dragooning of an unconvinced transcendent consciousness into the service of an immanent, limited enterprise like a human life.

VII

The final escape is suicide; but before adopting any hasty solutions, it would be wise to consider carefully whether the absurdity of our existence truly presents us with a *problem,* to which some solution must be found— a way of dealing with prima facie disaster. That is certainly the attitude with which Camus approaches the issue, and it gains support from the fact that we are all eager to escape from absurd situations on a smaller scale.

Camus—not on uniformly good grounds—rejects suicide and the other solutions he regards as escapist. What he recommends is defiance or scorn. We can salvage our dignity, he appears to believe, by shaking a fist at the world which is deaf to our pleas, and continuing to live in spite of it. This will not make our lives un-absurd, but it will lend them a certain nobility.[4]

This seems to me romantic and slightly self-pitying. Our absurdity warrants neither that much distress nor that much defiance. At the risk of falling into romanticism by a different route, I would argue that absurdity is one of the most human things about us: a manifestation of our most advanced and interesting characteristics. Like skepticism in epistemology, it is possible only because we possess a certain kind of insight—the capacity to transcend ourselves in thought.

If a sense of the absurd is a way of perceiving our true situation (even though the situation is not absurd until the perception arises), then what reason can we have to resent or escape it? Like the capacity for epistemological skepticism, it results from the ability to understand our human limitations. It need not be a matter for agony unless we make it so. Nor need it evoke a defiant contempt of fate that allows us to feel brave or proud. Such dramatics, even if carried on in private, betray a failure to appreciate the cosmic unimportance of the situation. If *sub specie aeternitatis* there is no reason to believe that anything matters, then that doesn't matter either, and we can approach our absurd lives with irony instead of heroism or despair.

4. "Sisyphus, proletarian of the gods, powerless and rebellious, knows the whole extent of his wretched condition: it is what he thinks of during his descent. The lucidity that was to constitute his torture at the same time crowns his victory. There is no fate that cannot be surmounted by scorn" (*The Myth of Sisyphus,* Vintage edition, p. 90).

E. D. KLEMKE

Living Without Appeal:
An Affirmative
Philosophy of Life[1]

From time to time, philosophers get together at congresses and symposia in which some philosophers read papers and others criticize and raise questions. To the layman, I am sure, the topics which are discussed seem highly technical and inaccessible, and the vocabulary used is, doubtless, unintelligible. Indeed, if the ordinary man were to drop in on such meetings, he would, I suspect, find the proceedings to be either totally incomprehensible or the occasion for howling laughter. To give some indication of what I am referring to, I shall list the titles of some recent philosophical papers, many of which are acknowledged to be very important works:

The meaning of a word
Performative—constative
Negative existentials
Excluders
Reference and referents
Proper names
On referring
Parenthetical verbs
Bare particulars
Elementarism, independence, and ontology
The problem of counterfactual conditionals

1. This paper was first read in the Last Lecture Series, at DePauw University, and was repeated, by request, three times. In a revised form, it was read as the Top Prof lecture at Roosevelt University. It was again revised for this volume.

Is existence a predicate?
Etc.

Upon hearing (or reading) papers such as these, the oridinary man would probably exclaim "What's this all got to do with philosophy?" And he would, no doubt, be in agreement with Kierkegaard, who once wrote:

> What the philosophers say about Reality is often as disappointing as a sign you see in a shop window which reads: Pressing Done Here. If you brought your clothes to be pressed, you would be fooled; for the sign is only for sale. (*Either/Or,* v. 1, p. 31.)

Now I have no quarrel with what goes on at these professional gatherings. I engage in such activities myself. I believe that most philosophical problems are highly technical and that the making of minute distinctions and the employment of a specialized vocabulary are essential for the solution of such problems. Philosophy here is in the same boat as any other discipline. For this reason, there is (and perhaps always will be) something aristocratic about the pursuit of philosophy, just as there is about the pursuit of theoretical physics or Peruvian excavation. The decriers of philosophy often overlook the fact that any discipline which amounts to more than a type of verbal diarrhea must proceed by making subtle distinctions, introducing technical terminology, and striving for as much rigor and precision as is possible. And the critics fail to see that, in philosophy as in other fields, by the very nature of the discipline, some problems will be somewhat rarified, and of interest mainly to the specialist.

On the other hand, I am inclined to think that the philosopher ought occasionally to leave the study, or the philosophical association lecture hall, or even the classroom, and, having shed his aristocratic garments, speak as a man among other men. For the philosopher is, after all, human too. Like other men, he eats, sleeps, makes love, drinks martinis (or perhaps cognac), gets the flu, files income tax, and even reads the newspapers. On such more democratic occasions, he ought to employ his analytical tools as diligently as ever. But he should select as his topic some issue which is of concern to all men, or at least most men, at some time in their lives. It is my hope that I have chosen such a topic for this essay.

The problem which I wish to discuss has been formulated in a single sentence by Camus (in *The Myth of Sisyphus*), which I take as a kind of "text." The sentence to which I am referring is: "Knowing whether or not one can live *without appeal* is all that interests me."[2] I say that I take this as a *kind* of text because, as so often, Camus overstates the point. Thus I

2. A. Camus, *The Myth of Sisyphus.* Tr. by J. O'Brien (New York: Vintage Books, 1959), p. 45.

would not—and perhaps most of us would not—say that, knowing whether
or not one can live without appeal is *all* that interests me. But I believe
that most of us would say that it certainly is one of those crucial problems
which each man must confront as he tries to make sense of his life in this
wondrously strange existence.

I

Prophets of doom and redemption seem to exist in almost every age, and
ours is no exception. It is commonly held by many present-day thinkers,
scholars, and poets, that the current state of the world and of many of the
individuals within it is one of disintegration and vacuity. As they see it:
Men of our age grope for disrupting principles and loyalties, and often reveal a destructive tension, a lack of wholeness, or an acute anxiety.
Whether or not this is a unique situation in history, as an account of the
present state of things, such disintegration is commonly mentioned. And
theorists in almost every discipline and pursuit have given analyses of the
current predicament and offered solutions. For example, philosophers and
theologians (Jaspers, Marcel, Swenson, Tillich, Schweitzer, Niebuhr),
scientists and scientific writers (Einstein, DeNuoy), sociologists (Sorokin),
historians (Butterfield), among others, have waved warning signs, sometimes in a last effort to "save civilization from utter destruction." I would
like to consider some points which are held in common by many of these
writers (and others whom I have not indicated) and then to comment
about those views. In this section, I shall state the common core of this
position. In the next section, I shall make my comments and show that
there is another genuine alternative.

According to many of the above writers (and others whom I have not
mentioned), our age is one in which a major catastrophe has taken place.
This has been designated as an increasing lack of a determining principle,
the severing of a determining bond, the loss of a determining passion, or
the rejection of a determining ultimate. What is the nature of this ultimate? It has been described as a principle by which finite forces are held
in equilibrium, a bond which relates all horizontally functioning powers
vertically to a realm beyond the finite. It is said to be a unifying and controlling power by which the varied inclinations, desires, and aims of an
individual may be kept in balance. It is characterized as an agency which
removes those oppositions and dichotomies which tend to destroy human
selfhood. It has been held, by writers such as the above, to be a *transcendent* and *unconditional ultimate,* the one indispensable factor for the attainment of a *meaningful* and *worthwhile* existence. In their view, in

order to prevent the destruction of individuals and cultures, and to provide a sense of direction and wholeness, the awareness of and relationship to such an ultimate are absolutely necessary.

Many of the writers have noted that, not only in intellectual circles, but at a much wider level, many individuals are increasingly refusing to accept the reality of this controlling ultimate. As they see it, such individuals have either remained content with a kind of vacuum in "the dimension of the spiritual," or they have "transvaluated and exalted immanent, finite forces" into a substitute for the transcendent. Men have tried—say these writers—to find equilibrium and unity through "natural," non-authoritative, self-regulating, temporal aims and principles, which they hold to be capable of an innate self-integration which requires no outside aid. According to these writers, this hope is futile, for as soon as reliance upon the transcendent ultimate ceases, disintegration results. Only when finite relationships, processes, and forces are referred back to a transhistorical order can integration, wholeness, meaning, and purpose be achieved. As long as men lack confidence in, or sever the bond to, the transcendent, their accomplishments and goals, no matter how noble or worthy, can have no final consistency or solidity. Rather, their efforts are mere remnants of an "atrophied world," shut up within the realm of immanence, intoxicated with itself, lured by "phantasms and idolatrous forces."

According to this view, the integrity of the individual is today threatened by the loss of belief in the transcendent ultimate and its replacement by a "devitalized" and "perverse" confidence in the all-sufficiency of the finite. The only remedy, we are told, is the recognition of the determining regulation of a dimension beyond the fleeting pace of the temporal world, by which alone existence can have worth and value.

At this point, one might be tempted to ask several questions of these writers:

(1) "Even if the above characterization of the world has some truth, must one look to transcendentalism as the remedy? Cannot a 'natural' philosophy or principle help us?"

The usual answer is: No. All naturalistic views reduce existence to mere finite centers and relationships. But all of these finite agencies are conditioned by others. All are therefore transitory and unstable. None can become a determining ultimate. Only a transcendent ultimate is capable of sustaining the kind of faith which gives human existence meaning and value.

(2) "But isn't this supernaturalism all over again? And doesn't it (as usual) imply either an unbridgeable gulf between the finite and the infinite or an external control or suppression of the finite by the so-called infinite?"

The customary reply is that this view may indeed be called supernaturalism. But (we are told) this does not imply the impossibility of any association of the finite and the infinite. For the ultimate, according to these writers, is not transcendent in the sense of being totally isolated from the finite, but, rather, is operative within the natural world. Furthermore (so the reply goes), the existence of a transcendent order does not entail either external control or suppression of the finite. It merely implies a human receptivity to a non-natural realm. That is, human achievement and value result from the impingement of the infinite upon the finite in moments of *kairos,* providing fullness and meaning but not at the price of denying the human activity which is involved. There always remains the awareness that the human subject is in a personal relation to another subject, a relation of supreme importance.

(3) "And how does one come to this relation?"

Perhaps mainly (say many of these writers) through suffering and sorrow, through a sense of sin and despair. When an individual sees that all finite centers and loyalties are fleeting and incapable of being lasting objects of faith, then he will renounce all previous efforts in despair, repent in humility, and gratefully make *the movement of faith* by which alone his life can become meaningful and worthwhile.

This, then, is the view which I propose to comment on. It is an all-or-nothing position. Its central thesis is that of a transcendent ultimate of absolute supremacy, which reigns over all finite things and powers, and *which alone is capable of providing meaning and worth to human existence.* Finite, historical centers can at best bring temporary assistance. They all wither with time and circumstances. Only when men turn from the finite to the infinite can they find (in the words of Kierkegaard) a hope and anticipation of the eternal which holds together all the "cleavages of existence."

II

I shall refer to the above view (which I have tried to portray justly) as transcendentalism. It contains three component theses. These are:

(1) There *exists* a transcendent being or ultimate with which man can enter into some sort of relation.

(2) Without such a transcendent ultimate, and the relation of faith to it, human life lacks *meaning, purpose,* and *integration.*

(3) Without such meaning or integration, human life is not *worthwhile.*

It is necessary to comment upon all three of these points.

(1) First, the thesis that there *exists* such a transcendent ultimate or power. I assume that those who assert the existence of a transcendent being intend their assertion to be a *cognitive* one. That is, they claim to be saying something which states a fact and which is capable of being either true or false. Thus they would not admit that their claim is merely an expression of feelings or attitudes. I also assume that those who make this assertion intend their statement to be interpreted *literally*. That is, they mean to say that the transcendent *really exists*. The transcendent presumably does not exist in the same sense in which Santa Claus may be said to "exist." These persons would, I assume, hold that the transcendent exists in actuality, although it may not exist in any empirical sense.

I ask: What *reasons* are there for holding that such an entity as the transcendent exists? I take it that I do not have to linger on such an answer as the testimony of a sacred book. The fact that the Bible or any other sacred writing asserts the existence of a transcendent is no more evidential to the existence of such a being than it is to the non-existence. All that a scriptural writing proves is that someone *believed* that a transcendent ultimate exists. And that is not at all the same as showing that such a being actually exists. The same may be said for the testimony of some unusual person—Moses, Jesus, Mohammed, etc. Furthermore, the fact that the testimony is made by a large number of persons does not substantiate the view. An impartial reading of history often shows that, on major issues, the majority is almost always wrong.

I also shall not linger on the traditional arguments for the existence of a god: The ontological, cosmological, teleological arguments, etc. Many theologians themselves now acknowledge that these are not so much arguments for the existence of such a being as they are explications of the affirmation of faith. Therefore, the fact that a certain segment of the universe is orderly, that it exhibits beauty, that it shows an adaptation of means to ends does not in any way provide evidence that there is one who orders, beautifies, and adapts.

Arguments from religious experience are also unconvincing. Due to their lack of intersubjective testability, the most that such arguments can demonstrate is that someone has had an unusual experience. They do not provide any evidence that the *object* of such an experience exists. That object may, of course, exist. But the occurrence of such an experience does not verify the existence of an actual, rather than imaginary, object. Suppose that, while a dentist is drilling my tooth, I have an experience of a blinding light or an unusual voice. I do not take this to be an adequate reason for saying that I *know* that I have now communed with the Absolute. I trust that you do not do so either.

What evidence, then, is there for the existence of a transcendent? I submit that there is *none*. And my reading of religious writings and my conversations with many of those who maintain the existence of the transcendent lead me to affirm that they also would agree that there is none. For they hold that the existence of the transcendent (although a cognitive claim) is apprehended, *not* in a cognitive relationship but in the relationship of *faith*.

Thus in the usual sense of the term "evidence," there seems to be no evidence for the existence of a transcendent ultimate. Why, then, should I accept such a claim? After all, throughout the rest of my philosophical activity *and* throughout my normal, everyday activities, I constantly rely upon criteria of evidence before accepting a cognitive claim. I emphasize that this holds for my *everyday* life and not merely for any philosophical or scientific beliefs which I may entertain. Not only do I accept or reject (say) the Principle of Rectilinear Propagation of Light because of evidence. I also ask for evidence in order to substantiate such simple claims as "The stylus in my stereo tone arm is defective," or "Jones eloped with his secretary."

It is clear that both believers and non-believers share this desire for evidence with me. At least, believers agree up to the point of the transcendent-claim. If I reject this claim because of lack of evidence, I do not think that I can be justly accused of being an extremist. Rather, I should be commended for my consistency!

The transcendentalist will reply: "But the usual criteria do not apply in this case. They work only for natural entities. The transcendent is not a natural being." I answer: Then the only reasonable procedure seems to be that of suspending my judgment, for I do not know of any non-natural criteria. The transcendentalist replies: "No, merely suspending your judgment implies that you think that some evidence might eventually be found. We are in a different dimension here. An act of *faith* is required."

I reply with two points: (a) In its normal usage, the term "faith" still implies evidence and reasons. Why do I have faith in Smith, but not in Jones? Obviously because of *reasons*. I do not have faith in people haphazardly and without evidence. (b) If I am told that faith in the transcendent is not faith in the normal sense, but a special act of commitment, then I can only honestly reply: *I have no need for such faith*. The transcendentalist retorts: "Ah, but you do, for only through faith in the transcendent can life have meaning; and surely you seek a life that is significant and worthwhile." And this leads us to the second thesis.

(2) The transcendentalist claims that without the transcendent and faith in the transcendent, human existence is without *meaning, purpose, integration*. Is this true? And if true in some sense, what follows?

(A) Let us take *meaning* first. Is there any reason to believe that without the existence of the transcendent, life has no meaning? That is, does the existence of meaning presuppose the existence of the transcendent?

It is necessary to distinguish between *objective* meaning and *subjective* meaning. An objective meaning, if there were such, would be one which is either structurally *part of* the universe, apart from human subjective evaluation; or dependent upon some *external agency* other than human evaluation. Two comments are in order: (i) *If* the notion of objective meaning is a plausible one, then I see no reason why it must be tied up with the existence of a transcendent being, for it certainly is not self-contradictory to hold that an objective meaning could conceivably exist even though a transcendent being did not. That is, the two concepts of "transcendent being" and "objective meaning" are not logically related in the way in which the two concepts "three" and "odd" (for example) are related. (ii) But, more fundamental, I find the notion of an objective meaning as difficult to accept as I do the notion of a transcendent being. Therefore I cannot rely upon the acceptance of objective meaning in order to substantiate the existence of the transcendent.

Further comment is needed on this point. It seems to me that there is no shred of evidence for the existence of an objective meaning in the universe. If I were to characterize the universe, attempting to give a complete description, I would do so in terms of matter in motion, or energy, or forces such as gravitation, or events, etc. Such a description is *neutral*. It can have no non-descriptive components. The same holds for a description of any segment of the universe. Kepler, for example, was entitled to say that the paths of the planets are elliptical, etc. But he was not entitled to say that this motion exhibits some fundamental, objective purpose more so than some other type of motion would. From the standpoint of present evidence, evaluational components such as meaning or purpose are not to be found in the universe as objective aspects of it. Such values are the result of human evaluation. With respect to them, we must say that the universe is valueless; it is *we* who evaluate, upon the basis of our subjective preferences. Hence, we do not discover values such as meaning to be inherent within the universe. Rather, we "impose" such values upon the universe.

When the transcendentalist holds that, without the transcendent, no objective meaning for human existence is possible, he assumes that the notion of an objective meaning is an intelligible one. But if one can show, as I believe one can, that the idea of objective meaning is an implausible one, then his argument has no point. In no way does it give even the slightest evidence for the existence of a transcendent ultimate.

However, it is possible that some transcendentalist would want to take a different position here. There are at least two alternatives which he might hold.

(i) The transcendentalist might *agree* that there is no *objective* meaning in the universe, that meaning *is* a function of human subjectivity. His point now is that *subjective* meaning is found if and only if there exists a transcendent. I reply with two points (1) This is a grandiose generalization, which might wow an imbecile but not anyone of normal intelligence, and, like most such generalizations, it is false. (I shall return to this point in connection with the transcendentalist's third thesis.) (2) The meaning which the transcendentalist here affirms cannot be subjective meaning, for it is dependent upon some external, non-human factor, namely, the existence of the transcendent. This sort of meaning is *not* a function of human subjectivity. Thus we are back where we were. The transcendentalist's views about meaning do not provide any evidence at all for the existence of a transcendent ultimate.

(ii) I mentioned that the transcendentalist may take a second alternative. He might want to hold: "Of course, the fact of meaning in human existence does not in any way prove, demonstratively or with probability, that there *is* a transcendent being. Therefore, I won't say that meaning in life is impossible unless the transcendent exists. I will merely say that one cannot find meaning unless *one has faith in* the transcendent. The fact of meaning testifies to the necessity of *faith*."

I reply again with two points. (1) This generalization is also false. I know of many humans who have found a meaningful existence without faith in the transcendent. (2) However, even if this statement were true—even if heretofore not a single human being had found meaning in his life without faith in the transcendent—*I should reject such meaning and search for some other kind.* To me, the price which the transcendentalist pays for his meaning is too dear. If I am to find any meaning in life, I must attempt to find it without the aid of crutches, illusory hopes, and incredulous beliefs and aspirations. I am perfectly willing to admit that *I may not find any meaning at all* (although I think I can, even if it is not of the noble variety of which the transcendentalist speaks). But at least I *must try* to find it on my own. And this much I know: I can strive for a meaning only if it is one which is within the range of my comprehension as an inquiring, rational *man*. A meaning which is tied to some transcendent entity— or to faith in such—is not intelligible to me. Again, I here maintain what I hold throughout the rest of my existence, both philosophically and simply as a living person. I can accept only what is comprehensible to me, i.e., that which is within the province of actual or possible experience, or that for which I find some sound reasons or evidence. Upon these grounds, I must reject any notion of meaning which is bound with the necessity of faith in some mysterious, utterly unknowable entity. If my life should turn out to be less happy thereby, then I shall have to endure it as such. As Shaw once said: "The fact that a believer is happier than a skeptic is no more to the point than the fact that a drunken man is happier than a sober one. The happiness of credulity is a cheap and dangerous quality."

(b) I shall not say much about the transcendentalist's claim that, without the transcendent, or without faith in it, human existence is *purposeless*. For if I were to reply in detail, I should do so in about the same manner as I did with respect to the matter of meaning. An objective purpose is as difficult to detect in the universe as an objective meaning. Hence, again, one cannot argue that there must be a transcendent or that faith in such is necessary.

(c) What about the transcendentalist's claim that, without the transcendent, or, without *faith* in the transcendent, no integration is possible?

(i) In one sense of the term, this assertion, too, is obviously false. There are many persons who have attained what might be called psychological integration, i.e., self-integration, integration of personality, etc., without faith in the transcendent. I know of dozens of people whose lives are integrated in this sense, yet have no transcendental commitments.

(ii) But perhaps the transcendentalist means something much more fundamental than this psychological thesis by his claim. Perhaps he is making some sort of metaphysical assertion—a statement about man and his place in the universe. Thus his assertion must be taken to mean that *metaphysical* integration is not achievable without the transcendent or without faith in it. Like Kierkegaard, he holds that the cleavages of existence cannot be held together without the transcendent. What shall we say to this interpretation?

I am not sure that I understand what such integration is supposed to be. But insofar as I do, it seems to me that it is not possible. I am willing to admit that, if such integration were achievable, it might perhaps be attained only by virtue of something transcendent. But I find no conclusive or even reasonable evidence that such integration has been achieved either by believers in the transcendent or by non-believers. Hence one cannot infer that there is a transcendent ultimate or that faith in such an entity is necessary.

What about the mystics? you ask. It would be silly for me to say that the mystics have not experienced something very unusual which they have *interpreted* as some sort of unity with the universe, or whatever it may be. They may, indeed, have *felt* that, at rare moments, they were "swallowed up in the infinite ocean of being," to quote James. But again, peculiar and non-intersubjectively testable experiences are not reliable evidence for any truth-claim. Besides, suppose that the mystics *had* occasionally achieved such unity with the universe. Still, this is somewhat irrelevant. For the point is, that *I,* and many beings like myself (perhaps most of you), have not been favored with such experiences. In fact, it appears that most people who have faith in the transcendent have not had such experiences. This is precisely *why* they have faith. If they had complete certainty, no faith would be needed. Thus faith itself does not seem to be enough for the achievement of integration; and if integration were obtained, faith would be unnecessary. Hence the transcendentalist's view that integration is achieved *via* faith in the transcendent is questionable.

But even if this last thesis were true, it does me no good. Once again, I cannot place my faith in an unknown X, in that which is incomprehensible to me. Hence I must accept the fact that, for me, life will remain without objective meaning, without purpose, and without metaphysical integration. *And I must go on from there.* Rather than crying for the moon, my task must be, as Camus said, to know whether or not one can live *without appeal.*

(3) This leads us to the transcendentalist's third (and most crucial) thesis: That without meaning, purpose, and integration, life is not *worthwhile.* From which he draws the conclusion that without a *transcendent* or *faith* in it, life is not worthwhile. I shall deal only with the claim that without *meaning,* life is not worthwhile. Similar comments could be made regarding purpose and integration.

If the transcendentalist's claim sounds plausible at all, it is only because he continues to confuse objective meaning with subjective meaning. It is true that life has no objective meaning. Let us face it once and for all. But from this it does not follow that life is not *worthwhile,* for it can still be subjectively meaningful. And, really, the latter is the only kind of meaning worth shouting about. An objective meaning—that is, one which is inherent within the universe or dependent upon external agencies— would, frankly, leave me cold. It would not be *mine.* It would be an outer, neutral thing, rather than an inner, dynamic achievement. I, for one, am *glad* that the universe has no meaning, for thereby is *man all the more glorious.* I willingly accept the fact that external meaning is non-existent (or if existent, certainly not apparent), for this leaves me free to *forge my own meaning.* The extent of my creativity and thereby my success in this undertaking depends partly on the richness of my own psyche. There are some persons whose subjectivity is poor and wretched. Once they give up the search for objective meaning, they may perhaps have a difficult time in finding life to be worthwhile. Such is the fate of the impoverished. But those whose subjectivity is enlarged—rationally, esthetically, sensually, passionally—may find life to be worthwhile by means of their creative activity of subjective evaluation, in which a neutral universe takes on color and light, darkness and shadow, becomes now a source of profound joy, now a cause for deep sorrow.

What are some ways by which such worthwhileness can be found? I can speak only for myself. I have found subjective meaning through such things as *knowledge, art, love, and work*. Even though I realize that complete and perfect knowledge of matters of fact is not attainable, this does not lessen my enthusiasm to know and to understand. Such pursuits may have no practical utility; they are not thereby any less significant. To know about the nature of necessary truth or the probable structure of the atom is intrinsically fascinating, to me. And what a wealth of material lies in the arts. A Bach fugue, a Vlaminck painting, a Dostoevsky novel; life is intensely enriched by things such as these. And one must not neglect mention of one's relationships of friendship and love. Fragmentary and imperfect as these often are, they nevertheless provide us with some of our most heightened moments of joy and value. Finally, of all of the ways which I listed, none is more significant and constantly sustaining to me than work. There have been times when I, like many others, no doubt, have suffered some tragedy which seemed unendurable. Every time, it has been my work that has pulled me through.

In short, even if life has no meaning, in an external, objective sense, this does not lead to the conclusion that it is not worth living, as the transcendentalist naively but dogmatically assumes. On the contrary, this fact opens up a greater field of almost infinite possibilities. For as long as I am *conscious*, I shall have the capacity with which to *endow* events, objects, persons, and achievements with value. Ultimately, it is through my *consciousness* and it alone that worth or value are obtained. Through consciousness, the scraping of horses' tails on cats' bowels (to use James' phrase) become the beautiful and melodic lines of a Beethoven string quartet. Through consciousness, a pile of rock can become the memorable Mount Alten which one has climbed and upon which one almost perished. Through consciousness, the arrangements of *P*s and *Q*s on paper can become the symbols of the formal beauty and certain truth of the realm of mathematical logic. Through consciousness, the gift of a carved little piece of wood, left at one's door by a friend, can become a priceless treasure. Yes, it is a *vital* and *sensitive consciousness* that counts. Thus there is a sense in which it is true, as many thinkers and artists have reminded us, that everything begins with my consciousness, and nothing has any worth except through my consciousness.

III

I shall conclude with an ancient story. "Once a man from Syria led a camel through the desert; but when he came to a dark abyss, the camel suddenly, with teeth showing and eyes protruding, pushed the unsuspecting paragon of the camel-driving profession into the pit. The clothes of the Syrian were caught by a rosebush, and he was held suspended over the pit, at the bottom of which an enormous dragon was waiting to swallow him. Moreover, two mice were busily engaged in chewing away the roots of the already sagging plant. Yet, in this desperate condition, the Syrian was thralled to the point of utmost contentment by a rose which adorned the bush and wafted its fragrance into his face."[3]

I find this parable most illuminating. We are all men hanging on the thread of a few rapidly vanishing years over the bottomless pit of death, destruction, and nothingness. Those objective facts are starkly real. Let us not try to disguise them. Yet I find it marvelously interesting that man's *consciousness*, his reason and his passion, can elevate these routine, objective, external events, in a moment of lucidity and feeling, to the status of a personally appropriated ideal—an ideal which does not annul those objective facts, but which *reinterprets* them and clothes them with the apparel of *man's subjectivity*.

It is time, once again, to speak personally. What your situation is, I cannot say. But I know that I am that Syrian, and that I am hanging over the pit. My doom is inevitable and swiftly approaching. If, in these few moments that are yet mine, I can find no rose to respond to, or rather, if I have lost the ability to respond, then I shall moan and curse my fate with a howl of bitter agony. But *if* I can, in these last moments, respond to a rose—or to a philosophical argument or theory of physics, or to a Scarlatti sonata, or to the touch of a human hand—I say, if I can so respond and can thereby transform an external and fatal event into a moment of conscious insight and significance, then I shall go down *without hope or appeal* yet *passionately triumphant and with joy*.

3. R. Hertz, *Chance and Symbol* (Chicago: University of Chicago, 1948), pp. 142–143. Another version of this parable appears in Tolstoy's *My Confession*.

PART III

QUESTIONING
THE QUESTION

❧ 13 ❧

KAI NIELSEN

Linguistic Philosophy
and ''The Meaning of Life''

I

Anglo-Saxon philosophy has in various degrees "gone linguistic." From the faithful attention to the niceties of plain English practiced by John Austin, to the use of descriptive linguistics initiated by Paul Ziff in his *Semantic Analysis,* to the deliberately more impressionistic concern with language typical of Isaiah Berlin and Stuart Hampshire, there is a pervasive emphasis by English-speaking philosophers on what can and cannot be said, on what is intelligible, and on what is nonsensical. When linguistic philosophy was first developing, many things were said to be nonsense which were not nonsense. However, this is something of the past, for linguistic philosophy has for a long time been less truculent and more diffident about what it makes sense to say, but only to become—some would say—unbelievably bland, dull and without a rationale that is of any general interest.[1]

Critics from many quarters have raised their voices to assault linguistic philosophy as useless pedantry remote from the perennial concerns of philosophy or the problems of belief and life that all men encounter when, in Hesse's terms, they feel to the full "the whole riddle of human destiny." Traditionally the philosophical enterprise sought, among other things, to give us some enlightenment about our human condition, but as philosophy "goes linguistic," it has traitorously and irresponsibly become simply talk

1. John Passmore remarks in his brief but thoroughly reliable and judicious *Philosophy in the Last Decade* (Sydney University Press: 1969) "Philosophy is once again cultivating areas it had declared wasteland, or had transferred without compunction to other owners," p. 5.

about the uses of talk. The philosopher has left his "high calling" to traffic in linguistic trivialities.

Criticism of linguistic philosophy has not always been this crude, but there has typically been at least the implied criticism that linguistic philosophy could not really do justice to the profound problems of men with which Plato, Spinoza or Nietzsche struggled.

It is my conviction that such a charge is unfounded. In linguistic philosophy there is a partially new technique but no "abdication of philosophy." Surely most linguistic philosophy is dull, as is most philosophy, as is most anything else. Excellence and insight in any field are rare. But at its best linguistic philosophy is not dull and it is not without point; furthermore, though it often is, it need not be remote from the concerns of men. It is this last claim—the claim that linguistic philosophy can have nothing of importance to say about the perplexities of belief and life that from time to time bedevil us—that I wish to challenge.

With reference to the concepts of human purpose, religion and the problematical notion "the meaning of Life," I want to show how in certain crucial respects linguistic philosophy can be relevant to the perplexities about life and conduct that reflective people actually face. "What is the meaning of Life?" has been a standby of both the pulpiteer and the mystagogue. It has not come in for extended analysis by linguistic philosophers, though Ayer, Wisdom, Baier, Edwards, Flew, Hepburn and Dilman have had some important things to say about this obscure notion which when we are in certain moods perplexes us all and indeed, as it did Tolstoy and Dostoevsky, may even be something that forces itself upon us in thoroughly human terms.[2] I want to show how the use of the analytical techniques of linguistic philosophy can help us in coming to grips with the problems of human purpose and the meaning of Life.

Part of the trouble centers around puzzles about the use of the word "meaning" in "What is the meaning of Life?" Since the turn of the century there has been a lot of talk in philosophical circles about "meanings" or "a meaning criterion" and a good measure of attention has been paid to considerations about the meanings of words and sentences. But the mark (token) "meaning" in "What is the meaning of Life?" has a very different use than it has in "What is the meaning of 'obscurantist'?" "What is the meaning of 'table'?" "What is the meaning of 'good'?" "What is the meaning of 'science'?" and "What is the meaning of 'meaning'?" In these other cases we are asking about the meaning or use of the word or words, and we are requesting either a definition of the word or an elucidation or de-

2. A. J. Ayer, "The Claims of Philosophy" in M. Natanson (ed.) *Philosophy of the Social Sciences* (Random House, 1963).

scription of the word's use. But in asking: "What is the meaning of Life?" we are not asking—or at least this is not our central perplexity—about "What is the meaning of the word 'Life'?" What then are we asking?

Indirection is the better course here. Consider some of the uses of the general formula: "What is the meaning of that?" How, in what contexts, and for what purposes does it get used? Sometimes we may simply not know the meaning of a word, as when we come across a word we do not understand and look it up in a dictionary or ask the person using it in conversation what it means. It is not that he is using the word in an odd sense and we want to know what *he means* by it, but that we want to know what is meant by that word as it is employed in the public domain.

There is the quite different situation in which it is not about words that we are puzzled but about someone's non-linguistic behavior. A friend gives us a dark look in the middle of a conversation in which several people are taking part and afterwards we ask him "What was the meaning of those dark looks?" We were aware when we noticed his dark look that he was disapproving of something we were doing but we did not and still do not know what. Our "What was the meaning of that?" serves to try to bring out what is the matter. Note that in a way here we are not even puzzled about the meaning of words. The recipient of the dark look may very well know he is being disapproved of; but he wants to know what for. Here "What is the meaning of that?" is a request for the point or the purpose of the action. In this way, as we shall see, it is closer to the question "What is the meaning of Life?" than questions about the meaning of a word or a sentence.

We also ask "What is the meaning of that?" when we want to know how a particular person on a particular occasion intends something. We want to know what *he means* by that. Thus if I say of some author that he writes "chocolate rabbit stories" you may well ask me what I mean by that. Here you are puzzled both about the meaning of the phrase "chocolate rabbit stories," for as with "the pine cone weeps" or "the rock cogitates" it is a deviant collection of words of indeterminate meanings, and about the point or purpose of making such a remark. After all, the point of making such an utterance may not be evident. Suppose I had said it to a stupid and pompous writer blown up with a false sense of his own importance. I could explain my meaning by saying that I was obliquely giving him to understand that his stories, like chocolate rabbits, were all out of the same mold: change the names and setting and you have the same old thing all over again. And the point of my utterance would also become evidence, i.e., to deflate the pompous windbag. The phrase "chocolate rabbit stories" has no fixed use in human discourse, but language is sufficiently elastic for me to be able to give it a use without generating any linguistic or conceptual shock. To explain my meaning I must make clear the use I am giving it and make evident why I choose to use such an odd phrase.

"What is the meaning of Life?" is in some very significant respects like this last question though it is of course also very different. It is different in being non-deviant and in being a profoundly important question in the way the other question clearly is not. But note the likeness. In the first place when we or other people ask this question we are often not at all sure what we are asking. In this practical context we may in a way even be puzzled about the word "life," though, as I have said, the question does not primarily function as a request for the explanation of the use of a word. There is a sense in which life does and there is a sense in which life does not begin and end in mystery. And when we ask about life here we are not asking Schrodinger's question or J. B. S. Haldane's. We are not in search of some property or set of properties that is common to and distinctive of all those things we call "living things." We are typically concerned with something very different and much vaguer. We are asking: "Is life just one damn thing after another until finally one day we die and start to rot? Or can I sum it up and find or at least give it some point after all? Or is this just a silly illusion born of fear and trembling?" These are desperately vague, amorphous questions, but—as Wisdom would surely and rightly say—not meaningless for all that. And for some of us, and perhaps for all of us, *sometimes,* they are haunting, edging questions, questions we agonize over, then evade, then again try to come to grips with.

First, I want to say that, like "What is the meaning of calling them chocolate rabbit stories?", "What is the meaning of Life?" does *not* have a clear use; but that it does not have a clear use does not, I repeat, entail or in any way establish that it does not have a use or even that it does not have a supremely important use.[3] Secondly, "What is the meaning of life?" most typically—though not always—functions as a request for the goals *worth* seeking in life though sometimes it may serve to ask if there are *any* goals worth seeking in life.[4] We are asking what (if anything) is the point to our lives? What (if anything) could give our lives purpose or point? In anguish we struggle to find the purpose, point or rationale of our grubby lives. But if this is the nature of the question, what would an answer look like? For this to be a fruitful question, all of us must ask ourselves indi-

3. John Wisdom has driven home this point with force. In particular see his "The Modes of Thought and the Logic of 'God'" in his *Paradox and Discovery* (California, 1965).

4. Ronald Hepburn has correctly stressed that this for some people may not be what is uppermost in their minds when they ask that question. See Hepburn's essay in this volume. See also Ilham Dilman's remarks about Hepburn's analysis in "Life and Meaning," *Philosophy,* 40 (October 1965).

vidually: what would we take as an answer? When we ask this we are apt to come up with a blank; and if we are readers of philosophical literature we may remember that, along with others, a philosopher as persuasive and influential as A. J. Ayer has said that all such questions are unanswerable. But if they are really unanswerable—or so it would seem—then they are hardly genuine questions.

I will concede that *in a sense* such questions are unanswerable, but in a much more important sense they *are* answerable. We can be intelligent about and reason about such questions. Any analysis which does not bring this out and elucidate it is confused and inadequate. In destroying pontifical pseudo-answers the baby has frequently gone down with the bath. In showing what kind of answers could not be answers to this question, the temptation is to stress that there are no answers at all and that indeed no answers are needed. I want to try to show why this is wrong and what an answer would look like.

II

How then is it possible for our life to have a meaning or a purpose? For a while, oddly enough, Ayer in his "The Claims of Philosophy" is a perfectly sound guide.[5] We do know what it is for a man to have a purpose. "It is a matter," Ayer remarks, "of his intending, on the basis of a given situation, to bring about some further situation which for some reason or other he conceives to be desirable."

But, Ayer asks, how is it possible for life *in general* to have a meaning or a purpose?

Well, there is one very simple answer. Life in general has a purpose if all living beings are tending toward a certain specifiable end. To understand the meaning of life or the purpose of existence it is only necessary to discover this end.

As Ayer makes perfectly clear, there are overwhelming difficulties with such an answer. In the first place there is no good reason to believe living beings are tending toward some specifiable end. But even if it were true that they are all tending toward this end such a discovery would not at all answer the question "What is the meaning or purpose of life?" This is so because when we human beings ask this exceedingly vague question

5. See Ayer, *op cit*. The rest of the references to Ayer in the text are from this essay. His brief remarks in his "What I Believe" in *What I Believe* (London: 1966) pp. 15–16 and in his introduction to *The Humanist Outlook*, A. J. Ayer (ed.), (London: 1968) pp. 6–7 are also relevant as further brief statements of his central claims about the meaning of life.

we are not just asking for *an explanation of* the facts of existence; we are asking for a *justification* of these facts. In asking this question we are seeking a way of life, trying as suffering, perplexed, and searching creatures to find what the existentialists like to call an "authentic existence." And as Ayer goes on to explain,

> a theory which informs them merely that the course of events is so arranged as to lead inevitably to a certain end does nothing to meet their need. For the end in question will not be one that they themselves have chosen. As far as they are concerned it will be entirely arbitrary; and it will be a no less arbitrary fact that their existence is such as necessarily to lead to its fulfillment. In short, from the point of view of justifying one's existence, there is no essential difference between a teleological explanation of events and a mechanical explanation. In either case, it is a matter of brute fact that events succeed one another in the ways they do and are explicable in the ways they are.

In the last analysis, an attempt to answer a question of why events are as they are must always resolve itself into saying only *how* they are. Every explanation of why people do such and such and why the world is so and so finally depends on a very general description. And even if it is the case, as Charles Taylor powerfully argues, that teleological explanations of human behavior are irreducible, Ayer's point here is not all weakened, for in explaining, teleologically or otherwise, we are still showing how things are; we are not justifying anything.[6]

When we ask: "What is the meaning of life?" we want an answer that is more than *just* an explanation or description of *how* people behave or *how* events are arranged or *how* the world is constituted. We are asking for a *justification* for our existence. We are asking for a justification for why life is as it is, and not even the most complete explanation and/or description of *how* things are ordered can answer this quite different question. The person who demands that some general description of man and his place in nature should entail a statement that man ought to live and die in a certain way is asking for something that can no more be the case than it can be the case that ice can gossip. To ask about the meaning of our lives involves asking how we should live, or whether any decision to live in one way is more *worthy* of acceptance than any other. Both of these questions are clearly questions of value; yet no statement of *fact* about how we in fact do live can by itself be sufficient to answer such questions. No statement of what ought to be the case can be deduced from a statement of what is the case. If we are demanding such an answer, then Ayer is perfectly right in claiming the question is unanswerable.

6. Charles Taylor, *The Explanation of Behavior* (Routledge and Kegan Paul, 1964).

Let me illustrate. Suppose, perhaps as a result of some personal crisis, I want to take stock of myself. As Kierkegaard would say, I want to appropriate, take to heart, the knowledge I have or can get about myself and my condition in order to arrive at some decision as to what sort of life would be most meaningful for me, would be the sort of life I would truly want to live if I could act rationally and were fully apprised of my true condition. I might say to myself, though certainly not to others, unless I was a bit of an exhibitionist, "Look Nielsen, you're a little bit on the vain side and you're arrogant to boot. And why do you gossip so and spend so much of your time reading science fiction? And why do you always say what you expect other people want you to say? You don't approve of that in others, do you? And why don't you listen more? And weren't you too quick with Jones and too indulgent with Smith?" In such a context I would put these questions and a host of questions like them to myself. And I might come up with some general explanations, good or bad, like "I act this way because I have some fairly pervasive insecurities." And to my further question, "Well, why do you have these insecurities?" I might dig up something out of my past such as "My parents died when I was two and I never had any real home." To explain why this made me insecure I might finally evoke a whole psychological theory and eventually perhaps even a biological and physiological theory, and these explanations about the nature of the human animal would themselves finally rest, in part at least, on various descriptions of how man does behave. In addition, I might, if I could afford it and were sufficiently bedevilled by these questions, find my way to a psychiatrist's couch and there, after the transference had taken place, I would eventually get more quite personalized explanations of my behavior and attitudes. But none of these things, in themselves, could tell me the meaning of life or even the meaning of my life, though they indeed might help me in this search. I might discover that I was insecure because I could never get over the wound of the loss of my father. I might discover that unconsciously I blamed myself. As a child I wished him dead and then he died so somehow I did it, really. And I would, of course, discover how unreasonable this is. I would come to understand that people generally react this way in those situations. In Tolstoy's phrase, we are all part of the "same old river." And, after rehearsing it, turning it over, taking it to heart, I might well gain control over it and eventually gain control over some of my insecurities. I could see and even live through again what *caused* me to be vain, arrogant and lazy. But suppose, that even after all these discoveries I really didn't want to change. After stocktaking, I found that I was willing to settle for the *status quo*. Now I gratefully acknowledge that this is very unlikely, but here we are concerned with the *logical* possibilities. "Yes, there are other ways of doing things," I say to myself, "but after all is said and done I have lived this way a long time and I would rather go on this way than change. This sort of life, is after all, the most meaningful one. This is how I really want to act and this is how I, and others like me, ought to act." What possible facts could anyone appeal to which would prove, in the sense of logically entail, that I was wrong and that the purpose of life or the meaning of life was very different than I thought it was? It is Ayer's contention, and I think he is right, that there are none.

"But you have left out God," someone might say. "You have neglected the possibility that there is a God and that God made man to His image and likeness and that God has a plan for man. Even Sartre, Heidegger and Camus agree that to ask 'What is the Meaning of Life?' or 'What is the purpose of human existence?' is, in effect, to raise the question of God. If there is a God your conclusion would not follow, and, as Father Copleston has said, if there is *no* God human existence can have no end or purpose other than that given by man himself."[7]

I would want to say, that the whole question of God or no God, Jesus or no Jesus, is entirely beside the point. Even if there were a God human existence can, in the relevant sense of "end," "purpose" or "meaning," have no other end, purpose or meaning than what we as human beings give it by our own deliberate choices and decisions.

Let us see how this is so. Let us suppose that everything happens as it does because God intends that it should. Let us even assume, as we in reality cannot, that we can know the purpose or intentions of God. Now, as Ayer points out, either God's "purpose is sovereign or it is not. If it is sovereign, that is, if everything that happens is necessarily in accordance with it, then it is true also of our behavior. Consequently, there is no point in our deciding to conform to it, for the simple reason that we cannot do otherwise." No matter what, we do God's purpose. There is no sense in saying it is *our* purpose, that it is something we have made our own by our own deliberate choice. I have not *discovered* a meaning for my life and other people have not *discovered* a meaning for their lives. If it were possible for us *not* to fulfill it, the purpose would not be God's *sovereign* purpose and if it is His sovereign purpose, it cannot, in the requisite sense, be *our* purpose, for it will not be something of which it would make sense to say that we chose it. It is just something that necessarily happens to us because of God's intentions. If we are compelled to do it, it is not *our* purpose. It is only our purpose if we want to do it and if we could have done otherwise.

On the other hand, if God's purpose is not sovereign and we are not

---

7. See his discussion of existentialism in his *Contemporary Philosophy*.

inexorably compelled to do what God wills, we have no reason to conform to God's purpose unless we independently judge it to be *good* or by our own indepedent decision make it our purpose. We cannot derive the statement "x is good" from "that Being whom people call 'God' says 'x is good' " or from "that Being whom people call 'God' wills x" unless we *independently* judge that whatever this Being *says* is good *is good* or whatever that Being wills *ought* to be done. Again, as Ayer remarks, this "means that the significance of our behavior depends finally upon our own judgments of value; and the concurrence of a diety then becomes superfluous."[8]

The basic difficulty, as Ayer makes clear, is that in trying to answer the questions as we have above, we have really misunderstood the question. "What-is-the-meaning-of-that?" and "What-is-the-purpose-of-that?" questions can be very different. We have already noted some of the differences among "What-is-the-meaning-of-that?" questions, and we have seen that "What is the meaning of Life?" in many contexts at least can well be treated as a "What-is-the-purpose-of-that?" question. But "What is the purpose of life?" is only very superficially like "What is the purpose of a blotter?" "What is the purpose of brain surgery?" or "What is the purpose of the liver?" The first is a question about a human artifact and in terms of certain assumed ends we can say quite explicitly, independently of whether or not we want blotters, what the purpose of blotters is. Similarly brain surgery is a well-known human activity and has a well-known rationale. Even if we are Christian Scientists and disapprove of surgery altogether, we can understand and agree on what the purpose of brain surgery is, just as we all can say Fearless Fosdick is a good safecracker, even though we disapprove of safecrackers. And again, in terms of the total functioning of the human animal we can say what livers are for, even though the liver is not an artifact like a blotter. If there is a God and God made man, we *might* say the question "What is the purpose of human life?" is very like "What is the purpose of umbrellas?" The human animal then becomes a Divine artifact. But, even if all this were so, we would not—as we have already seen—have an answer to the *justificatory* question we started with when we asked, "What is the meaning of life?" If we knew God's purpose for man, we would know what man was made for. But we would not have an answer to our question about the meaning of life, for we would not know if there was purpose *in* our lives or if we could find a point in acting one way rather than another. We would only know that there was some-

8. While I completely agree with the central thrust of Ayer's argument here, he has, I believe, overstated his case. Even if our behaviour finally depends on our own standards of value, it does not follow that the concurrence of the deity, if there is one, is superfluous, for we could still find crucial moral guidance from our grasp of something of God's wisdom.

thing—which may or may not be of value—that we were constructed, "cut out," to be.

Similarly, if an Aristotelian philosophy is correct, "What is the purpose of life?" would become very like "What is the purpose of the liver?" But here again a discovery of what end man is as a matter of fact tending toward would not answer the perplexity we started from, that is to say, it would not answer the question, "What is the meaning of life, how should men live and die?" We would only learn that "What is the purpose of life?" could admit of two very different uses. As far as I can see, there are no good reasons to believe either that there is a God or that the human animal has been ordered for some general end; but even if this were so it would not give us an answer to the question: "What is the meaning of life?"

This is so because the question has been radically misconstrued. When we ask: "What is the meaning of life?" or "What is the purpose of human existence?" we are normally asking, as I have already said, questions of the following types: "What should we seek?" "What ends—if any—are worthy of attainment?" Questions of this sort require a very different answer than any answer to: "What is the meaning of 'obscurantism'?" "What is the purpose of the ink-blotter?" and "What is the purpose of the liver?" Ayer is right when he says: "what is required by those who seek to know the purpose of their existence is not a factual description of the way that people actually do conduct themselves, but rather a decision as to how they *should* conduct themselves." Again he is correct in remarking: "There is—a sense in which it can be said that life does have a meaning. It has for each of us whatever meaning we severally *choose* to give it. The purpose of a man's existence is constituted by the ends to which he, consciously or unconsciously, devotes himself."

Ayer links this with another crucial logical point, a point which the existentialists have dramatized as some kind of worrisome "moral discovery." Ayer points out that "in the last resort . . . each individual has the responsibility of making the choice of how he ought to live and die" and that it is logically impossible that someone else, in some authoritative position, can make that choice for him. If someone gives me moral advice in the nature of the case I must decide whether or not to follow his advice, so again the choice is finally my own. This is true because moral questions are primarily questions about what to do. In asking how I ought to live, I am trying to make up my mind how to act. And to say I deliberately acted in a certain way implies that I decided to do it. There is no avoiding personal choice in considering such questions.

But Ayer, still writing in the tradition of logical empiricism, often writes as if it followed from the truth of what we have said so far, that there could be no reasoning about "How ought man to live?" or "What is the meaning of life?" Thus Ayer says at one point in "The Claims of Philosophy": "He [the moral agent] cannot prove his judgments of value are correct, for the simple reason that no judgment of value is capable of proof." He goes on to argue that people have no way of demonstrating that one judgment of value is superior to another. A decision between people in moral disagreement is a "subject for persuasion and finally a matter of individual choice."

As we have just seen there is a sound point to Ayer's stress on choice vis-a-vis morality, but taken as a whole his remarks are at best misleading. There is reasoning about moral questions and there are arguments and proofs in morality. There are principles in accordance with which we appraise our actions, and there are more general principles, like the principle of utility or the principles of distributive justice in accordance with which we test our lower-level moral rules. And there is a sense of "being reasonable" which, as Hume and Westermarck were well aware, has distinctive application to moral judgments. Thus, if I say, "I ought to be relieved of my duties, I'm just too ill to go on" I not only must believe I am in fact ill, I must also be prepared to say, of any of my colleagues or anyone else similarly placed, that in like circumstances they too ought to be relieved of their duties if they fall ill. There is a certain *generality* about moral discourse and a man is not reasoning morally or "being reasonable" if he will not allow those inferences. Similarly, if I say "I want x" or "I prefer x" I need not, though I may, be prepared to give reason why I want it or prefer it, but if I say "x is the right thing to do" or "x is good" or "I ought to do x" or "x is worthy of attainment," I must—perhaps with the exception of judgments of intrinsic goodness—be prepared to give *reasons* for saying "x is the right thing to do," "x is good," "I ought to do x" and the like. (Note, this remark has the status of what Wittgenstein would call a grammatical remark.)

It is indeed true in morals and in reasoning about human conduct generally that justification must come to an end; but this is also true in logic, science and in common sense empirical reasoning about matters of fact; but it is also true that the end point in reasoning over good and evil is different than in science and the like, for in reasoning about how to act, our judgment finally terminates in a choice—a decision of principle. And here is the truth in Ayer's remark that moral judgments are "*finally* a matter of individual choice." But, unless we are to mislead, we must put the emphasis on "finally," for a dispassionate, neutral analysis of the uses of the language of human conduct will show, as I have indicated, that there is reasoning, and in a relevant sense, "objective reasoning," about moral questions. It is not at all a matter of pure persuasion or goading someone into sharing your attitudes.

I cannot, of course, even begin to display the full range of the reasoning which has sought to establish this point. But I hope I have said enough to block the misleading implications of Ayer's otherwise very fine analysis. Early linguistic philosophy was primarily interested in (1) the descriptive and explanatory discourse of the sciences, and (2) in logico-mathematico discourse; the rest was rather carelessly labelled, "expressive or emotive discourse." But the thrust of the work of linguistic philosophers since the Second World War has corrected that mistaken emphasis, as recent analytical writing in ethics makes evident. Here I commend to you R. M. Hare's *The Language of Morals,* and his *Freedom and Reason,* Stephen Toulmin's *An Examination of the Place of Reason in Ethics,* Kurt Baier's *The Moral Point of View,* Marcus Singer's *Generalization in Ethics,* P. H. Nowell-Smith's *Ethics,* Bernard Mayo's *Ethics and the Moral Life,* or George von Wright's *The Varieties of Goodness.* They would also reinforce the point I tried briefly to make against Ayer, as would an examination of the essays of Philippa Foot or John Rawls.[9]

III

There are, however, other considerations that may be in our minds when we ask "What is the meaning of life?" or "Does life have a meaning?'" In asking such questions, we may *not* be asking "What should we seek?" or "What goals are worth seeking really?" Instead we may be asking "Is *anything* worth seeking?" "Does it matter finally what we do?" Here, some may feel, we finally meet the real tormenting "riddle of human existence."

Such a question is not simply a moral question: it is a question concerning human conduct, a question about how to live one's life or about whether to continue to live one's life. Yet when we consider what an answer would look like here we draw a blank. If someone says "Is anything worthwhile?" we gape. We want to reply: "Why, sitting in the sunshine in the mornings, seeing the full moon rise, meeting a close friend one hasn't seen in a long time, sleeping comfortably after a tiring day, all these things and a million more are most assuredly worthwhile. Any life devoid of experiences of this sort would most certainly be impoverished."

Yet this reply is so obvious we feel that something different must be

9. I have discussed these issues in my "Problems of Ethics" and "History of Contemporary Ethics," both in Vol. 3 of *The Encyclopedia of Philosophy,* Paul Edwards, ed., (Macmillan, 1967).

intended by the questioner. The questioner knows that we, and most probably he, ordinarily regard such things as worthwhile, but he is asking if these things or *anything* is worthwhile *really*? These things *seem* worthwhile but are they in reality? And here we indeed do not know what to say. If someone queries whether it is really worthwhile leaving New York and going to the beach in August we have some idea of what to say; there are some criteria which will enable us to make at least a controversial answer to this question. But when it is asked, in a philosophical manner, *if anything, ever* is really worthwhile, it is not clear that we have a genuine question before us. The question borrows its form from more garden-variety questions but when we ask it in this general way do we actually know what we mean? If someone draws a line on the blackboard, a question over the line's straightness can arise only if some criterion for a line's being straight is accepted. Similarly only if some criterion of worthiness is accepted can we intelligibly ask if a specific thing or anything is worthy of attainment.

But if a sensitive and reflective person asks, "Is anything worthwhile, really?" could he not be asking this because, (1) he has a certain vision of human excellence, and (2) his austere criteria for what is worthwhile have developed in terms of that vision? Armed with such criteria, he might find nothing that man can in fact attain under his present and foreseeable circumstances *worthy* of attainment. Considerations of this sort seem to be the sort of considerations that led Tolstoy and Schopenhauer to come to such pessimistic views about life. Such a person would be one of those few people, who as one of Hesse's characters remarks, "demand the utmost of life and yet cannot come to terms with its stupidity and crudeness." In terms of his ideal of human excellence nothing is worthy of attainment.

To this, it is natural to respond, "If this is our major problem about the meaning of life, then this is indeed no intellectual or philosophical riddle about human destiny. We need not like Steppenwolf return to our lodging lonely and disconsolate because life's 'glassy essence' remains forever hidden, for we can well envisage, in making such a judgment, what would be worthwhile. We can say what a meaningful life would look like even though we can't attain it. If such is the question, there is no 'riddle of human existence', though there is a pathos to human life and there is the social-political pattern problem of how to bring the requisite human order into existence. Yet only if we have a conception of what human life should be can we feel such pathos."

If it is said in response to this that what would really be worthwhile could not possibly be attained, an absurdity has been uttered. To say something is worthy of attainment implies that, everything else being equal, it ought to be attained. But to say that something ought to be attained implies that it *can* be attained. Thus we *cannot* intelligibly say that something is worthy of attainment but that it cannot possibly be attained. So in asking "Is anything worthy of attainment?" we must acknowledge that there are evaluative criteria operative which guarantee that what is sincerely said to be worthy of attainment is at least in principle attainable. And as we have seen in speaking of morality, "x is worthy of attainment" does not mean "x is preferred," though again, in asserting that something is worthy of attainment, or worthwhile, we imply that we would choose it, everything else being equal, in preference to something else. But we cannot intelligibly speak of a choice if there is no possibility of doing one thing rather than another.

Life is often hard and, practically speaking, the ideals we set our hearts on, those to which we most deeply commit ourselves, may in actual fact be impossible to achieve. A sensitive person may have an ideal of conduct, an ideal of life, that he assents to without reservation. But the facts of human living being what they are, he knows full well that this ideal cannot be realized. His ideals are intelligible enough, logically their achievement is quite possible, but as a matter of *brute fact* his ideals are beyond his attainment. If this is so, is it worthwhile for him and others like him to go on living or to strive for anything at all? Can life, under such circumstances, be anything more than an ugly habit? For such a man, "What is the meaning of life?" has the force of "What *point* can a life such as mine have under these circumstances?" And in asking whether such a life has a point he is asking the very question we put above, viz. can life be worth living under such conditions.

Again such a question is perfectly intelligible and is in no way unanswerable any more than any other question about how to act, though here too we must realize that the facts of human living *cannot* be sufficient for a man simply to read off an answer without it in any way affecting his life. Here, too, *any* answer will require a decision or some kind of effective involvement on the part of the person involved. A philosopher can be of help here in showing what kind of answers we cannot give, but it is far less obvious that he can provide us with a set of principles that together with empirical facts about his condition and prospects, will enable the perplexed man to know what he ought to do. The philosopher or any thoughtful person who sees just what is involved in the question can give some helpful advice. Still the person involved must work out an answer in anguish and soreness of heart.

However, I should remind him that no matter how bad his own life was, there would always remain something he could do to help alleviate the sum total of human suffering. This certainly has value and if he so oriented his life, he could not say that his life was without point. I would also argue that in normal circumstances he could not be sure that his ideals of life would permanently be frustrated, and if he held ideals that would be badly frustrated under almost any circumstances, I would get him to look again at his ideals. Could such ideals really be adequate? Surely man's reach must exceed his grasp, but how far should we go? Should not any ideal worth its salt come into some closer involvement with the realities of human living? And if one deliberately and with self-understanding plays the role of a Don Quixote can one justifiably complain that one's ideals are not realized? Finally, it does not seem to me reasonable to expect that *all* circumstances can have sufficient meaning to make them worthwhile. Under certain circumstances life is not worth living. As a philosopher, I would point out this possibility and block those philosophical-religious claims that would try to show that this could not possibly be.

Many men who feel the barbs of constant frustration, come to feel that their ideals have turned out to be impossible, and ask in anguish—as a consequence—"Does life really have any meaning?" To a man in such anguish I would say all I have said above and much more, though I am painfully aware that such an approach may seem cold and unfeeling. I know that these matters deeply affect us; indeed they can even come to obsess us, and when we are so involved it is hard to be patient with talk about what can and cannot be said. But we need to understand these matters as well; and, after all, what more can be done along this line than to make quite plain what is involved in his question and try to exhibit a range of rational attitudes that could be taken toward it, perhaps stressing the point that though Dr. Rieux lost his wife and his best friend, his life, as he fought the plague, was certainly not without point either for him or for others. But I would also try to make clear that finally an answer to such a question must involve a decision or the having or adopting of a certain attitude on the part of the person involved. This certainly should be stressed and it should be stressed that the question "Is such a life meaningful?" is a sensible question, which admits of a non-obscurantist, non-metaphysical treatment.

IV

There are many choices we must make in our lives and some choices are more worthwhile than others, though the criteria for what is worthwhile are in large measure at least context-dependent. "It's worthwhile going to Leningrad to see the Hermitage" is perfectly intelligible to someone who knows and cares about art. Whether such a trip to Leningrad is worthwhile for such people can be determined by these people by a visit to the Museum. "It's worthwhile fishing the upper Mainistee" is in exactly the same category, though the criteria for worthwhileness are not the same. Such statements are most assuredly perfectly intelligible; and no adequate grounds have been given to give us reason to think that we should philosophically tinker with the ordinary criteria of "good art museum" or "good trout fishing." And why should we deny that these and other things are really worthwhile? To say "Nothing is worthwhile since all pales and worse still, all is vain because man must die" is to mistakenly assume that because an eternity of even the best trout fishing would be not just a bore but a real chore, that trout fishing is therefore not worthwhile. Death and the fact (if it is a fact) that there is nothing new under the sun need not make all vanity. That something must come to an end can make it all the more precious: to know that love is an old tale does not take the bloom from your beloved's cheek.

Yet some crave a more general answer to "Is anything worthwhile?" This some would say, is what they are after when they ask about the meaning of life.

As I indicated, the criteria for what is worthwhile are surely in large measure context-dependent, but let us see what more we can say about this need for a more general answer.

In asking "Why is anything worthwhile?" if the "why" is a request for *causes,* a more general answer can be given. The answer is that people have preferences, enjoy, admire and approve of certain things and they can and sometimes do reflect. Because of this they find some things worthwhile. This, of course, is not what "being worthwhile" *means,* but if people did not have these capacities they would not find anything worthwhile. But *reasons* why certain things are worthwhile are dependent on the thing in question.

If people find x worthwhile they generally prefer x, approve of x, enjoy x, or admire x on reflection. If people did not prefer, approve of, enjoy or admire things then nothing would be found to be worthwhile. If they did not have these feelings the notion of "being worthwhile" would have no role to play in human life; but it does have a role to play and, as in morality, justification of what is worthwhile must finally come to an end with the reflective choices we make.

Moral principles, indeed, have a special onerousness about them. If something is a moral obligation, it is something we ought to do through and through. It for most people at least and from a moral point of view for everyone overrides (but does not exhaust) all non-moral considerations about what is worthwhile. If we are moral agents and we are faced with the necessity of choosing either A or B, where A, though very worthwhile, is a non-moral end and where B is a moral one, we must choose B. The force of the "must" here is logical. From a moral point of view there is no alternative but to choose B. Yet we do not escape the necessity of decision for we still must *agree* to *adopt* a moral point of view, to *try* to act as moral agents. Here, too, we must finally make a decision of principle.[10] There are good Hobbesian reasons for adopting the moral point of view but if one finally would really prefer "a state of nature" in which all were turned against all, rather than a life in which there was a freedom from this and at least a minimum of cooperation between human beings, then these reasons for adopting the moral point of view would not be compelling to such a person. There is, in the last analysis, no escape from making a choice.

In asking "What is the meaning of Life?" we have seen how this question is in reality a question concerning human conduct. It asks either "What should we seek?" or "What ends (if any) are really worthwhile?" I have tried to show in what general ways such questions are answerable. We can give reasons for our moral judgments and moral principles and the whole activity of morality can be seen to have a point, but not all questions concerning what is worthwhile are moral questions. Where moral questions do not enter we must make a decision about what, on reflection, we are going to seek. We must ascertain what—all things considered—really answers to our interests or, where there is no question of anything answering to our interests or failing to answer to our interests, we should decide what on reflection we prefer. What do we really want, wish to approve of, or admire? To ask "Is anything worthwhile?" involves our asking "Is there nothing that we, on reflection, upon knowledge of ourselves and others, want, approve of, or admire?" When we say "So-and-so is worthwhile" we are making a normative judgment that cannot be derived from determining what we desire, admire or approve of. That is to say, these statements do not entail statements to the effect that so and so is worthwhile. But in determining what is worthwhile this is finally all we have to go on. In saying something is worthwhile, we (1) *express* our preference, admiration or approval; (2) in some sense imply that we are prepared to defend our choice with *reasons;* and (3) in effect, indicate our belief that others like us in the relevant respects and similarly placed, will find it worthwhile too. And the answer to our question is that, of course, there are things we humans desire, prefer, approve of, or admire. This being so, our question is not unanswerable. Again we need not fly to a metaphysical enchanter.

---

10. I have discussed the central issues involved here at length in my "Why Should I Be Moral?" *Methods,* 15 (1963).

As I said, "Is anything really worthwhile, really worth seeking?" makes us gape. And "atomistic analyses," like the one I have just given, often leave us with a vague but persistent feeling of dissatisfaction, even when we cannot clearly articulate the grounds of our dissatisfaction. "The real question," we want to say, "has slipped away from us amidst the host of distinctions and analogies. We've not touched the deep heart of the matter at all."

Surely, I have not exhausted the question for, literally speaking, it is not one question but a cluster of loosely related questions all concerning "the human condition"—how man is to act and how he is to live his life even in the face of the bitterest trials and disappointments. Questions here are diverse, and a philosopher, or anyone else, becomes merely pretentious and silly when he tries to come up with some formula that will solve, resolve or dissolve the perplexities of human living. But I have indicated in skeletal fashion how we can approach general questions about "What (if anything) is worth seeking?" And I have tried to show how such questions are neither meaningless nor questions calling for esoteric answers.

V

We are not out of the woods yet. Suppose someone were to say: "Okay, you've convinced me. Some things are worthwhile and there is a more or less distinct mode of reasoning called moral reasoning and there are canons of validity distinctive of this *sui generis* type reasoning. People do reason in the ways that you have described, but it still remains the case that here one's attitudes and final choices are relevant in a way that it isn't necessarily the case in science or an arguments over plain matters of fact. But when I ask: 'How ought men act?' 'What is the meaning of life?' and 'What is the meaning of *my* life?, how should I live and die?' I want an answer that is logically independent of any human choice or any pro-attitude toward any course of action or any state of affairs. Only if I can have that kind of warrant for my moral judgments and ways-of-life will I be satisfied."

If a man demands this and continues to demand this after dialectical examination we must finally leave him unsatisfied. As linguistic philosophers there is nothing further we can say to him. In dialectical examination we can again point out to him that he is asking for the logically impossible, but if he recognizes this and persists in asking for that which is impossible there are no further rational arguments that we can use to establish our point. But, prior to this last-ditch stand, there are still some things that we can say. We can, in detail and with care, point out to him, describe fully for him, the rationale of the moral distinctions we do make and the functions of moral discourse. A full description here will usually break this kind of obsessive perplexity. Furthermore, we can make the move Stephen Toulmin makes in the last part of his *The Place of Reason in Ethics*. We can describe for him another use of "Why" that Toulmin has well described as a "limiting question."[11]

Let me briefly explain what this is and how it could be relevant. When we ask a "limiting question" we are not really asking a question at all. We are in a kind of "land of shadows" where there are no clear-cut uses of discourse. If we just look at their grammatical form, "limiting questions" do not appear to be extra-rational in form, but in their depth grammar—their actual function—they clearly are. "What holds the universe up?" looks very much like "What holds the Christmas tree up?" but the former, in common sense contexts at least, is a limiting question while the latter usually admits of a perfectly obvious answer. As Toulmin himself puts it, limiting questions are "questions expressed in a form borrowed from a familiar mode of reasoning, but not doing the job which they normally do within that mode of reasoning."[12] A direct answer to a limiting question never satisfies the questioners. Attempted "answers" only regenerate the question, though often a small change in the questions themselves or their context will make them straightforward questions. Furthermore, there is no standard interpretation for limiting questions sanctioned in our language. And limiting questions do not present us with any genuine alternatives from which to choose.

Now "limiting questions" get used in two main contexts. Sometimes, they merely express what Ryle, rather misleadingly, called a "category mistake." Thus someone who was learning English might ask: "How hot is blue?" or "Where is anywhere?" And, even a native speaker of English might ask as a *moral* agent, "Why ought I to do what is right?" We "answer" such questions by pointing out that blue cannot be hot, anywhere is not a particular place, and that if something is indeed right, this entails that it ought to be done. Our remarks here are grammatical remarks, though our speaking in the material mode may hide this. And if the questioner's "limiting question" merely signifies that a category mistake has been made, when this is pointed out to the questioner, there is an end to the matter. But more typically and more interestingly, limiting questions do not *just* or at all indicate category mistakes but express, as well or independently, a *personal predicament*. Limiting questions may express anxiety, fear, hysterical apprehensiveness about the future, hope, de-

11. Stephen Toulmin, *An Examination of the Place of Reason in Ethics* (Cambridge University Press, 1950).
12. *Ibid.*, p. 205.

spair, and any number of attitudes. Toulmin beautifully illustrates from
the writings of Dostoevsky an actual, on-the-spot use, of limiting questions:

> He was driving somewhere in the steppes. . . . Not far off was a
> village, he could see the black huts, and half the huts were burnt down,
> there were only the charred beams sticking out. As they drove in, there
> were peasant women drawn up along the road . . .
> "Why are they crying? Why are they crying?" Mitya [Dmitri] asked,
> as they dashed gaily by.
> "It's the babe," answered the driver, "the babe is weeping."
> And Mitya was struck by his saying, in his peasant way, "the babe,"
> and he liked the peasant's calling it a "babe." There seemed more pity
> in it.
> "But why is it weeping?" Mitya persisted stupidly. "Why are its
> little arms bare? Why don't they wrap it up?"
> "The babe's cold, its little clothes are frozen and don't warm it."
> "But why is it? Why?" foolish Mitya still persisted.
> "Why, they're poor people, burnt out. They've no bread. They're
> begging because they've been burnt out."
> "No, no," Mitya, as it were still did not understand. "Tell me why
> it is those poor mothers stand there? Why are people poor? Why is the
> babe poor? Why is the steppe barren? Why don't they hug each other and
> kiss? Why don't they sing songs of joy? Why are they so dark from black
> misery? Why don't they feed the babe?"
> And he felt that, though his questions were unreasonable, and sense-
> less, yet he wanted to ask just that, and he had to ask it just in that way.
> And he felt that a passion of pity, such as he had never known before, was
> rising in his heart, that he wanted to cry, that he wanted to do something
> for them all, so that the babe should weep no more, so that the dark-faced,
> dried-up mother should not weep, that no one should shed tears again
> from that moment . . .
> "I've had a good dream, gentlemen," he said in a strange voice, with
> a new light, as of joy, in his face.[13]

It is clear that we need not, may not, from the point of view of
analysis, condemn these uses of language as illicit. We can point out that
it is a muddle to confuse such questions with literal questions, and that
such questions have no fixed *literal* meaning, and that as a result there are
and can be no fixed literal ways of answering them, but they are indeed,
genuine uses of language, and not the harum-scarum dreams of undisci-
plined metaphysics. When existentialist philosophers and theologians state
them as profound questions about an alleged ontological realm there is
room for complaint, but as we see them operating in the passage I quoted

---

13. *Ibid.*, p. 210.

from *The Brothers Karamazov,* they seem to be not only linguistically proper but also an extremely important form of discourse. It is a shame and a fraud when philosophers "sing songs" as a substitute for the hard work of philosophizing, but only a damn fool would exclude song-singing, literal or metaphorical, from the life of reason, or look down on it as a somehow inferior activity. Non-literal "answers" to these non-literal, figurative questions, when they actually express personal predicaments or indeed more general human predicaments may, in a motivational sense, *goad* people to do one thing or another that they *know* they ought to do or they may comfort them or give them hope in time of turmoil and anxiety. I am not saying this is their only use or that they have no other respectable rationale. I do not at all think that; but I am saying that here is a rationale that even the most hard nosed positivist should acknowledge.

The man who demands "a more objective answer" to his question, "How ought men to live?" or "What is the meaning of Life?" may not be just muddled. If he is *just* making a "category mistake" and this is pointed out to him, he will desist, but if he persists, his limiting question probably expresses some anxiety. In demanding an answer to an evaluative question that can be answered independently of any attitudes he might have or choices he might make, he may be unconsciously expressing his fear of making decisions, his insecurity and confusion about what he really wants, and his desperate desire to have a Father who would make all these decisions for him. And it is well in such a context to bring Weston LaBarre's astute psychological observation to mind. "Values," LaBarre said, "must from emotional necessity be viewed as absolute by those who use values as compulsive defenses against reality, rather than properly as tools for the exploration of reality."[14] This remark, coming from a Freudian anthropologist, has unfortunately a rather metaphysical ring, but it can be easily enough de-mythologized. The point is, that someone who persists in these questions, persists in a demand for a totally different and "deeper" justification or answer to the question "What is the meaning of Life?" than the answer that such a question admits of, may be just expressing his own insecurity. The heart of rationalism is often irrational. At such a point the only reasoning that will be effective with him, if indeed any reasoning will be effective with him, may be psychoanalytic reasoning. And by then, of course you have left the philosopher and indeed all questions of justification far behind. But again the philosopher can describe the kinds of questions we can ask and the point of these questions. Without advocating anything at all he can make clearer to us the structure of "the life of reason" and the goals we human beings do prize.

14. Weston LaBarre, *The Human Animal* (University of Chicago, 1954).

VI

There is another move that might be made in asking about this haunting question: "What is the meaning of Life?" Suppose someone were to say: "Yes I see about these 'limiting questions' and I see that moral reasoning and reasoning about human conduct generally are limited modes of reasoning with distinctive criteria of their own. If I am willing to be guided by reason and I can be reasonable there are some answers I can find to the question: 'What is the meaning of Life?' I'm aware that they are not cut and dried and that they are not simple and that they are not even by any means altogether the same for all men, but there are some reasonable answers and touchstones all the same. You and I are in perfect accord on that. But there is one thing I don't see at all, 'Why ought I to be guided by reason anyway?' and if you cannot answer this for me I don't see why I should think that your answer—or rather your schema for an answer—about the meaning of Life is, after all, really any good. It all depends on how you *feel,* finally. There are really no answers here."

But again we have a muddle; let me very briefly indicate why. If someone asks: "Why ought I to be guided by reason anyway?" or "Is it really good to be reasonable?" one is tempted to take such a question as a paradigm case of a "limiting question," and a very silly one at that. But as some people like to remind us—without any very clear sense of what they are reminding us of—reason has been challenged. It is something we should return to, be wary of, realize the limits of, or avoid, as the case may be. It will hardly do to take such a short way with the question and rack it up as a category mistake.

In some particular contexts, with some particular people, it is (to be paradoxical, for a moment) reasonable to question whether we ought to follow reason. Thus, if I am a stubborn, penny-pinching old compulsive and I finally take my wife to the "big-city" for a holiday, it might be well to say to me: "Go on, forget how much the damn tickets cost, buy them anyway. Go on, take a cab even if you can't afford it." But to give or heed such advice clearly is not, in any fundamental sense, to fly in the face of reason, for on a deeper level—the facts of human living being what they are—we are being guided by reason.

It also makes sense to ask, as people like D. H. Lawrence press us to ask, if it really pays to be reasonable. Is the reasonable, clear-thinking, clear-visioned, intellectual animal really the happiest, in the long run? And can his life be as rich, as intense, as creative as the life of Lawrence's sort of man? From Socrates to Freud it has been assumed, for the most part, that self-knowledge, knowledge of our world, and rationality will bring happiness, if anything will. But is this really so? The whole Socratic tradition may be wrong at this point. Nor is it obviously true that the reasonable man, the man who sees life clearly and without evasion, will be able to live the richest, the most intense or the most creative life. I hope these things are compatible but they may not be. A too clear understanding may dull emotional involvement. Clear-sightedness may work against the kind of creative intensity that we find in a Lawrence, a Wolfe or a Dylan Thomas.

But to ask such questions is not in a large sense to refuse to be guided by reason. Theoretically, further knowledge could give us at least some vague answers to such unsettling questions; and, depending on what we learned and what decisions we would be willing to make, we would then know what to do. But clearly, we are not yet flying in the face of reason, refusing to be guided by reason at all. We are still playing the game according to the ground rules of reason.

What is this question, "Why should I be guided by reason?" or "Why be reasonable?" if it isn't any of these questions we have just discussed? If we ask this question and take it in a very general way, the question is a limiting one and it does involve a category mistake. What could be *meant* by asking: "Why ought we *ever* use reason at all?" That to ask this question is to commit a logical blunder, is well brought out by Paul Taylor when he says:

> . . . it is a question which would never be asked by anyone who thought about what he was saying, since the question, to speak loosely, answers itself. It is admitted that no amount of arguing in the world can make a person who does not want to be reasonable want to be. For to argue would be to give reasons, and to give reasons already assumes that the person to whom you give them is *seeking* reasons. That is it assumes he is reasonable. A person who did not want to be reasonable in any sense would never ask the question, "Why be reasonable?" For in asking the question, Why? he is seeking reasons, that is, he is being reasonable in asking the question. The question calls for the use of reason to justify *any* use of reason, including the use of reason to answer the question.[15]

In other words, to ask the question, as well as answer it, commits one to the use of reason. To ask: "Why be guided by reason at all?" is to ask "Why be reasonable, ever?" As Taylor puts it, "The questioner is thus seeking good reasons for seeking good reasons," and this surely is an absurdity. Anything that would be a satisfactory answer would be a "tautol-

15. Paul Taylor, "Four Types of Relativism," *Philosophical Review* (1956).

ogy to the effect that it is reasonable to be reasonable. A negative answer to the question, Is it reasonable to be reasonable? would express a self-contradiction."

If all this is pointed out to someone and he still persists in asking the question in this logically senseless way there is nothing a philosopher *qua* philosopher can do for him, though a recognition of the use of limiting questions in discourse may make this behavior less surprising to the philosopher himself. He might give him all five volumes of *The Life of Reason* or *Vanity Fair* and say, "Here, read this, maybe you will come to see things differently." The philosopher himself might even sing a little song in praise of reason, but there would be nothing further that he could say to him, philosophically: but by now we have come a very long way.

VII

Ronald Hepburn is perceptive in speaking of the conceptual "darkness around the meaning-of-life questions."[16] We have already seen some of the reasons for this; most generally, we should remark here that people are not always asking the same question and are not always satisfied by answers of the same scope when they wrestle with meaning-of-life questions. And often, of course, the questioner has no tolerably clear idea of what he is trying to ask. He may have a strong gut reaction about the quality and character of his own life and the life around him without the understanding or ability to conceptualize why he feels the way he does. Faced with this situation, I have tried to chart some of the contexts in which "What is the meaning of Life?" is a coherent question and some of the contexts in which it is not. But there are some further contexts in which "meaning-of-life questions" get asked which I have not examined.

There are philosophers who will agree with me that in a world of people with needs and wants already formed, it can be shown that life in a certain "subjective sense" has meaning, but they will retort that this is not really the central consideration. What is of crucial importance is whether we can show that the universe is better with human life than without it. If this cannot be established then we cannot have good reason to believe that life really has meaning, though in the subjective senses we have discussed, we can still continue to say it has meaning.[17]

16. Ronald W. Hepburn, "Questions About the Meaning of Life."
17. See in this context Hans Reiner, *Der Sinn unseres Daseins* (Tubingen; 1960). This view has been effectively criticized by Paul Edwards, "Meaning and Value of Life," *The Encyclopedia of Philosophy*, Paul Edwards, ed. (Macmillan, 1967), Vol. 4, pp. 474–476.

If we try to answer this question, we are indeed brought up short, for we are utterly at a loss about what it would be like to ascertain whether it is better for the universe to have human life than no life at all. We may have certain attitudes here but no idea of what it would be like to know or have any reason at all to believe that "It is better that there is life" is either true or false or reasonably asserted or denied. It is quite unlike "It is better to be dead than to live with such a tumor." Concerning this last example, people may disagree about its correctness, but they have some idea of what considerations are relevant to settling the dispute. But with "It is better that there be life" we are at a loss.

We will naturally be led into believing that "What is the meaning of Life?" is an unanswerable question reflecting "the mystery of existence," if we believe that to answer that question satisfactorily we will have to be able to establish that it is better that there is life on earth than no life at all. What needs to be resisted is the very acceptance of that way of posing the problem. We do not need to establish that it is better that the universe contains human life than not in order to establish that there is a meaning to life. A life without purpose, a life devoid of satisfaction and an alienated life in which people are not being true to themselves is a meaningless life. The opposite sort of life is a meaningful or significant life. We have some idea of the conditions which must obtain for this to be so, i.e. for a man's life to have significance. We are not lost in an imponderable mystery here and we do not have to answer the question of whether it is better that there be human life at all to answer that question. Moreover, this standard non-metaphysical reading of "What is the meaning of Life?" is no less objective than the metaphysical reading we have been considering. There are no good grounds at all for claiming that this metaphysical "question" is the real and objective consideration in "What is the meaning of Life?" and that the more terrestrial interpretations I have been considering are more subjective. This transcendental metaphysical way of stating the problem utilizes unwittingly and without justification arbitrary *persuasive* definitions of "subjective" and "objective." And no other grounds have been given for *not* sticking with the terrestrial readings.

A deeper criticism of the account I have given of purpose and the meaning of life is given by Ronald Hepburn.[18] It is indeed true that life cannot be meaningful without being purposeful in the quite terrestrial sense I have set out, but, as Hepburn shows, it can be purposeful and still be meaningless.

---

18. Hepburn's criticisms are directed toward an earlier version of this essay, "Linguistic Philosophy and 'The Meaning of Life'," *Cross-Currents*, 14 (Summer 1964).

One may fill one's days with honest, useful and charitable deeds, not
doubting them to be of value, but without feeling that these give one's
life meaning or purpose. It may be profoundly boring. To seek meaning
is not just a matter of seeking justification for one's policies, but of trying
to discover how to organise one's vital resources and energies around these
policies. To find meaning is not a matter of judging these to be worthy,
but of seeing their pursuit as in some sense a fulfillment, as involving self-
realisation as opposed to self-violation, and as no less opposed to the performance of a dreary task.[19]

A person's life can have significance even when he does not realize it and
even when it is an almost intolerable drudge to him, though for human
life generally to have significance this could not almost invariably be true
for the human animal. But one's own life could not have significance *for
oneself* if it were such a burden to one. To be meaningful to one, one's
life must be purposive *and* it must be a life that the liver of that life finds
satisfactory in the living of it. These conditions sometimes obtain and
when it is also true that some reasonable measure of an individual's purposive activity adds to the enhancement of human life, we can say that
his life is not only meaningful to him but meaningful *sans phrase*.[20]

This is still not the end of the matter in the struggle to gain a sense
of the meaning of life, for, as Hepburn also points out, some will not be
satisfied with a purely terrestrial and non-metaphysical account of the
type I have given of "the meaning of Life."[21] They will claim "that life
could be thought of as having meaning only so long as that meaning was
believed to be a matter for discovery, not for creation and value-decision."[22]
They will go on to claim that "to be meaningful, life would have to be
comprehensively meaningful and its meaning invulnerable to assault.
Worthwhile objectives must be ultimately realisable despite appearances."[23]

However, even if they are not satisfied with my more piecemeal and
terrestrial facing of questions concerning the meaning of life, it does not
follow that life can only have meaning if it has meaning in the more comprehensive and less contingent way they seek. It may be true that life will
only have meaning *for them* if these conditions are met, but this does not
establish that life will thus lack meaning unless these conditions are met.

19. *Ibid.*
20. *Ibid.*
21. Questions About the Meaning of Life." For arguments of this type see F. C.
Copleston, "Man and Metaphysics I," *The Heythrop Journal*, I, 2 (January 1960),
p. 16. See in addition his continuation of this article in successive issues of *The
Heythrop Review* and his *Positivism and Metaphysics* (Lisbon: 1965).
22. "Questions About the Meaning of Life."
23. *Ibid.*

That is to say, it may be found significant by the vast majority of people, including most non-evasive and reflective people, when such conditions are met and it may be the case that everyone *should* find life meaningful under such conditions.

It is not the case that there is some general formula in virtue of which we can say what the meaning of life is, but it still remains true that men can through their purposive activity give their lives meaning and indeed find meaning in life in the living of it. The man with a metaphysical or theological craving will seek "higher standards" than the terrestrial standards I have utilized.

Is it rational to assent to that craving, to demand such "higher standards," if life is really to be meaningful? I want to say both "Yes" and "No."

On the one hand, the answer should be "No," if the claim remains that for life to be meaningful at all it must be comprehensively meaningful. Even without such a comprehensive conception of things there can be joy in life, morally, aesthetically and technically worthwhile activity and a sense of human purpose and community. This is sufficient to give meaning to life. And as Ayer perceptively argues and as I argued earlier in the essay, and as Hepburn argues himself, the man with a metaphysical craving of the transcendental sort will not be able to succeed in finding justification or rationale for claims concerning the significance of life that is any more *authoritative* and any more certain or invulnerable to assault than the non-metaphysical type rationale I have adumbrated. In actuality, as we have seen, such a comprehensive account, committed, as it must be, to problematic transcendental metaphysical and/or theological conceptions, is more vulnerable than my purely humanistic reading of this conception.

On the other hand, the answer should be "Yes" if the claim is reduced to one asserting that to try to articulate a comprehensive picture of human life is a desirable thing. However, it should be noted that this is quite a reduction in claim. In attempting to make such an articulation, the most crucial thing is not to wrestle with theological considerations about the contingency of the world or eternal life, but to articulate a comprehensive normative social and political philosophy in accordance with which we could set forth at least some of the conditions of a non-alienated life not simply for a privileged few but for mankind generally. We need to show in some general manner what such a life would look like and we need to attempt again, and with a reference to contemporary conditions, what Marx so profoundly attempted, namely, to set out the conditions that could transform our inegalitarian, unjust, vulgar and—as in countries such as South Africa and the United States—brutal capitalist societies into truly human societies.[24] Linguistic philosophers and bourgeois philosophers generally have been of little help here, though the clarity they have inculcated into philosophical work and into political and moral argument will be a vital tool in this crucial and yet to be done task.[25] When this task is done, if it is done, then we will have the appropriate comprehensive picture we need, and it is something to be done without any involvement with theology, speculative cosmology or transcendental metaphysics at all.[26]

24. For a contemporary Marxist account see Adam Schaff, *A Philosophy of Man* (London: 1963). But also note the criticism of Schaff's views by Christopher Hollis in "What is the Purpose of Life?" *The Listener, 70* (1961), pp. 133–136.
25. The strength and limitations here of linguistic analysis as it has been practiced are well exhibited in Ayer's little essay "Philosophy and Politics."
26. If what I have argued above is so, many of the esoteric issues raised by Milton Munitz in his *The Mystery of Existence* and in his contribution to *Language, Belief, and Metaphysics*, Kiefer and Munitz, eds. (New York: 1970) can be bypassed.

☙ 14 ❧

JOHN WISDOM

The Meanings of
the Questions of Life

When one asks "What is the meaning of life?" one begins to wonder
whether this large, hazy and bewildering question itself has any meaning. Some people indeed have said boldly that the question has no meaning. I believe this is a mistake. But it is a mistake which is not without
excuse. And I hope that by examining the excuse we may begin to remedy
the mistake, and so come to see that whether or not life has a meaning it
is not senseless to enquire whether it has or not. First, then, what has led
some people to think that the whole enquiry is senseless?

There is an old story which runs something like this: A child asked
an old man "What holds up the world? What holds up all things?" The
old man answered "A giant." The child asked "And what holds up the
giant? You must tell me what holds up the giant." The old man answered
"An elephant." The child said, "And what holds up the elephant?" The
old man answered "A tortoise." The child said "You still have not told me
what holds up all things. For what holds up the tortoise?" The old man
answered "Run away and don't ask me so many questions."

From this story we can see how it may happen that a question which
looks very like sensible meaningful questions may turn out to be a senseless, meaningless one. Again and again when we ask "What supports
this?" it is possible to give a sensible answer. For instance what supports
the top-most card in a house of cards? The cards beneath it which are in
their turn supported by the cards beneath them. What supports all the
cards? The table. What supports the table? The floor and the earth. But
the question "What supports all things, absolutely all things?" is different.
It is absurd, it is senseless, like the question "What is bigger than the
largest thing in the world?" And it is easy to see why the question "What

supports all things?" is absurd. Whenever we ask, "What supports thing A or these things A, B, C," then we can answer this question only by mentioning some thing other than the thing A or things A, B, C about which we asked "What supports it or them." We must if we are to answer the question mention something D other than those things which form the subject of our question, and we must say that this thing is what supports them. If we mean by the phrase "all things" absolutely all things which exist then obviously there is nothing outside that about which we are now asked "What supports all this?" Consequently any answer to the question will be self-contradictory just as any answer to the question "What is bigger than the biggest of all things" must be self-contradictory. Such questions are absurd, or, if you like, silly and senseless.

In a like way again and again when we ask "What is the meaning of this?" we answer in terms of something other than this. For instance imagine that there has been a quarrel in the street. One man is hitting another man on the jaw. A policeman hurries up. "Now then" he says, "what is the meaning of all this?" He wants to know what led up to the quarrel, what caused it. It is no good saying to the policeman "It's a quarrel." He knows there is a quarrel. What he wants to know is what went before the quarrel, what led up to it. To answer him we must mention something other than the quarrel itself. Again suppose a man is driving a motor car and sees in front of him a road sign, perhaps a red flag, perhaps a skull and cross bones. "What does this mean?" he asks and when he asks this he wants to know what the sign points to. To answer we must mention something other than the sign itself, such as a dangerous corner in the road. Imagine a doctor sees an extraordinary rash on the face of his patient. He is astonished and murmurs to himself "What is the meaning of this?" He wants to know what caused the strange symptoms, or what they will lead to, or both. In any case in order to answer his question he must find something which went before or comes after and lies outside that about which he asks "What does this mean?" This need to look before or after in order to answer a question of the sort "What is the meaning of this?" is so common, so characteristic, a feature of such questions that it is natural to think that when it is impossible to answer such a question in this way then the question has no sense. Now what happens when we ask "What is the meaning of life?"

Perhaps someone here replies, the meaning, the significance of this present life, this life on earth, lies in a life hereafter, a life in heaven. All right. But imagine that some persistent enquirer asks, "But what I am asking is what is the meaning of all life, life here and life beyond, life now and life hereafter? What is the meaning of all things in earth and heaven?" Are we to say that this question is absurd because there cannot be anything beyond all things while at the same time any answer to "What is the meaning of all things?" must point to some thing beyond all things?

Imagine that we come into a theatre after a play has started and are obliged to leave before it ends. We may then be puzzled by the part of the play that we are able to see. We may ask "What does it mean?" In this case we want to know what went before and what came after in order to understand the part we saw. But sometimes even when we have seen and heard a play from the beginning to the end we are still puzzled and still ask what does the whole thing mean. In this case we are not asking what came before or what came after, we are not asking about anything outside the play itself. We are, if you like, asking a very different sort of question from that we usually put with the words "What does this mean?" But we are still asking a real question, we are still asking a question which has sense and is not absurd. For our words express a wish to grasp the character, the significance of the whole play. They are a confession that we have not yet done this and they are a request for help in doing it. Is the play a tragedy, a comedy or a tale told by an idiot? The pattern of it is so complex, so bewildering, our grasp of it still so inadequate, that we don't know what to say, still less whether to call it good or bad. But this question is not senseless.

In the same way when we ask "What is the meaning of all things?" we are not asking a senseless question. In this case, of course, we have not witnessed the whole play, we have only an idea in outline of what went before and what will come after that small part of history which we witness. But with the words "What is the meaning of it all?" we are trying to find the order in the drama of Time. The question may be beyond us. A child may be able to understand, to grasp a simple play and be unable to understand and grasp a play more complex and more subtle. We do not say on this account that when he asks of the larger more complex play "What does it mean?" then his question is senseless, nor even that it is senseless for him. He has asked and even answered such a question in simpler cases, he knows the sort of effort, the sort of movement of the mind which such a question calls for, and we do not say that a question is meaningless to him merely because he is not yet able to carry out quite successfully the movement of that sort which is needed in order to answer a complex question of that sort. We do not say that a question in mathematics which is at present rather beyond us is meaningless to us. We know the type of procedure it calls for and may make efforts which bring us nearer and nearer to an answer. We are able to find the meaning which lies not outside but within very complex but still limited wholes whether these are dramas of art or of real life. When we ask "What is the

meaning of all things?" we are bewildered and have not that grasp of the
order of things the desire for which we express when we ask that question. But this does not render the question senseless nor make it impossible
for us to move towards an answer.

We must however remember that what one calls answering such a
question is not giving an answer. I mean we cannot answer such a question in the form: "The meaning is this."

Such an idea about what form answering a question must take may
lead to a new despair in which we feel we cannot do anything in the way
of answering such a question as "What is the meaning in it all?" merely
because we are not able to sum up our results in a phrase or formula.

When we ask what is the meaning of this play or this picture we
cannot express the understanding which this question may lead to in the
form of a list of just those things in the play or the picture which give it
its meaning. No. The meaning eludes such a list. This does not mean that
words quite fail us. They may yet help us provided that we do not expect
of them more than they can do.

A person who is asked what he finds so hateful or so lovable in another may with words help himself and us in grasping what it is that so
moves him. But he will only mislead us and himself if he pretends that
his words are a complete account of all that there is in the matter.

It is the same when we ask what is it in all things that makes it all
so good, so bad, so grand, so contemptible. We must not anticipate that
the answer can be given in a word or in a neat list. But this does not mean
that we can do nothing towards answering these questions nor even that
words will not help us. Indeed surely the historians, the scientists, the
prophets, the dramatists and the poets have said much which will help
any man who asks himself: Is the drama of time meaningless as a tale
told by an idiot? Or is it not meaningless? And if it is not meaningless is
it a comedy or a tragedy, a triumph or a disaster, or is it a mixture in
which sweet and bitter are for ever mixed?

❧ 15 ❧

R. W. HEPBURN

Questions about the Meaning of Life

Claims about "the meaning of life" have tended to be made and discussed in conjunction with bold metaphysical and theological affirmations. For life to have meaning, there must (it is assumed) be a comprehensive divine plan to give it meaning, or there must be an intelligible cosmic process with a "telos" that a man needs to know if his life is to be meaningfully orientated. Or, it is thought to be a condition of the meaningfulness of life, that values should be ultimately "conserved" in some way, that no evil should be unredeemable and irrational. And it may be claimed that if death were to end our experience, meaninglessness would triumph.

Because of this rich metaphysical background, the agnostic or naturalist faces a problem when he asks himself, "Does life have meaning, for me?": or more formally, "Can the vocabulary of life as 'meaningful' or 'meaningless' still play a role in my naturalistic interpretation of things?" The answer is not simple; for the informal logic of the vocabulary itself is not simple. This is a situation in which the naturalistic philosopher is tempted in one of two directions: either to renounce the vocabulary—as too deeply entangled with unacceptable beliefs, or to radically redefine the terms "meaning," "meaningful" and their cognates, thereby giving them work to do, but misleadingly different from their traditional work. The task for such a philosopher may be to mediate delicately between these poles, and in the course of trying to do that a good many complexities in the language of "meaning" may come to light—complexities of interest perhaps not only to the naturalistic philosopher himself.

In the last few years some analytical philosophers have in fact written on the expression "the meaning of life"; for instance, Kurt Baier, Inaugural Lecture, "The Meaning of Life" (Canberra, 1957), Antony Flew,

"Tolstoi and the Meaning of Life," *Ethics* (Jan. 1963), Kai Nielsen, "Linguistic Philosophy and 'The Meaning of Life'," *Cross-Currents* (Summer, 1964). Their analyses overlap at many points, though not completely. I shall first of all offer a very brief, compressed and somewhat schematised account of these analyses, conflating them where they overlap, and then discuss some of the issues they raise.[1]

According to the interpretations being now worked out, questions about the meaning of life are, very often, conceptually obscure and confused. They are amalgams of logically diverse questions, some coherent and answerable, some neither. A life is not a statement, and cannot therefore have linguistic meaning. But admittedly we do use the word "meaning" outside linguistic contexts. We speak of the meaning of a gesture, of a transaction, of a disposition of troops; and in such cases we are speaking of the point or purpose or end of an act or set of acts. This usage suggests an equation between meaningfulness and purposiveness. For a life to be meaningful, it must be purposeful: or—to make life meaningful is to pursue valuable ends.

To adopt this schema, however, already involves a shift from traditional ways of speaking about life as meaningful. Meaning is not now something to be found, as awaiting discovery, but is imparted to it by the subject himself. A person looks in vain for meaning and is needlessly frustrated when he cannot find it—if he conceives it as somehow existing prior to his decisions about what policies to pursue.

To say that "making life meaningful" is a matter of "pursuing valuable, worthwhile ends" is to say that it is an activity that indispensably involves value judgments. The description of cosmic patterns, tendencies or trends does not obviate the need to make autonomous judgments about the worthwhileness or otherwise of following, or promoting or opposing any of these. This is true no less of statements about God or a hereafter. Since claims about God and hereafter are ultimately claims about what is the case, not what ought to be, no conclusion will follow deductively from them about values. Even in the religious context, the question (as now analysed) still arises, "What ends are *worthy* of pursuit?" Religious propositions cannot guarantee meaningfulness. Conversely, it is argued, loss of religious or metaphysical belief does not entail the denial of meaningfulness.

If we concentrate on the question "What is the purpose of life?" rather than on "What is its meaning?" we are still dealing with the ques-

1. I had completed this article before seeing relevant studies by John Wisdom— *Paradox and Discovery* (1966), Ch. IV, and by Ilham Dilman—*Philosophy*, Oct. 1965.

tions "What ends shall I choose?" "What purposes shall I hold to?" The analytical philosopher characteristically and quite intentionally changes the question from singular to plural, from "purpose" to "purposes." He will claim that the original question contains a presupposition that must be rejected, the presupposition that life, if purposeful, can have only a single purpose, that only a single policy is worthwhile. Why may not a purposeful existence be a network of many purposes, with trajectories of varying reach, scope and seriousness? "We are not and cannot always be doing or caring about one big thing."[2]

The phrase, "the purpose of human life," can be offensive on a second count, because it may suggest an analogy with the purpose of an instrument, utensil, tool or organ in a living body. The theist, who does take this analogy seriously, sees human life (with its help) as subordinated to the intentions, activities of another being, who "assigns" (Baier) tasks and roles to men, and thereby imbues his life with purpose. Two senses of "purpose" must be contrasted: the first, the sense in which one has a purpose if one "purposes" or plans to do something, perhaps with the help of an artefact, an instrument, or other means: the second is the sense in which an artefact itself has a purpose, that is, a function. Only the first of these senses is compatible with moral autonomy, with being an autonomous purposer. To subordinate oneself wholly to the purposes of another is to forfeit moral status. This amounts to a moral argument against the manner in which theism seeks to secure purpose for human life. We must resist the temptation to translate "What is the purpose of life?" into "What are people *for?*"

We have seen two ways in which recent analyses of questions about the meaning of life seek to disconnect those questions from metaphysical and religious claims, or at least try to deprive such claims of any unique, privileged importance in the handling of the questions. Religious and metaphysical statements are still statements of fact, and therefore logically cannot in themselves be answers to questions about meaning. Second: if human life is given purpose by virtue of man's fulfilling the task assigned him by God, it will be "purpose" in the autonomy-denying, dignity-destroying sense. Two further arguments tend to the same general conclusion.

Consider the familiar claim that life is meaningless if death ends all, that a necessary condition of life being meaningful is immortality or resurrection. Against this it is argued that there is no entailment between temporal finiteness and disvalue, futility. We can and do love flowers that

---

2. H. D. Aiken, *Reason and Conduct* (New York, 1962), p. 374. See chapter xvi as a whole.

fade; and the knowledge that they will fade may even enhance their pre-
ciousness. To be everlasting, that is, is no necessary or sufficient condition
of value and worthwhileness, nor therefore of meaningfulness. An eternity
of futility is not logically impossible.

Again, the quest for the meaning of life is very often thought of, but
thought of confusedly, as a quest for *esoteric wisdom*, metaphysical or
theological. On this, Flew's discussion of Tolstoi is interesting and relevant. First, we find in Tolstoi's *A Confession* (O.U.P. ed., 1940) an account of a period of "arrest of life," the loss of a sense that life has any
meaning or purpose, an inner deadness and disorientation. Tolstoi asked,
"What is it for?" "What does it lead to?" His reflection took various pessimistic turns, including notably a refusal to account anything worthwhile,
if death ends all. It struck him eventually, however, that the vanity and
brevity of life were well enough known to very simple and unreflective
people: yet in their cases there was seldom any "arrest of life." Tolstoi
concluded that these people must know the meaning of life, the meaning
that eluded the learned Tolstoi himself: it must be, not a piece of rational
knowledge, but some non-rational or supra-rational knowledge to which
they somehow had access. Flew now contests the assumption that these
people necessarily had any esoteric, mystical knowledge that Tolstoi lacked.
"What we surely need here is Ryle's distinction between knowing *how*
and knowing *that*; the peasants may indeed know how to live their lives
free of all sophisticated psychological disabilities, but this by no means
presupposes the possession of any theoretical knowledge" hidden from
such a man as Tolstoi. In fact, the characters in Tolstoi's novels know
better. In *War and Peace* Pierre's "mental change" is a coming to have (a
learning how to have) peace of mind, not a matter of acquiring new information, new dogma. The same is true of Levin and Hadji Murad. In
so far as this kind of discovery of meaning is religious, it is so in a way
that may be "analysable in terms of ethics and psychology only." Crucially,
to know the meaning of life is to know how to live, as at one stage Tolstoi
did not.

If we continue thus and develop these last themes, it becomes clear that
the first-mentioned schematic account of the meaning of life is quite misleadingly oversimple. To give life meaning cannot be just a matter of
pursuing worthy projects, for that account fails to cope with phenomena
like Tolstoi's arrest of life—or John Stuart Mill's during his mental crisis
of 1826. More generally, it is quite possible to make various value-
judgments in cold blood, while yet suffering from a sense of meaninglessness. One may fill one's days with honest, useful and charitable deeds,
not doubting them to be of value, but without feeling that these give

one's life meaning or purpose. It may be profoundly boring. To seek meaning is not just a matter of seeking justification for one's policies, but of trying to discover how to organise one's vital resources and energies around these policies. To find meaning is not a matter of judging these to be worthy, but of seeing their pursuit as in some sense a fulfilment, as involving self-realisation as opposed to self-violation, and as no less opposed to the performance of a dreary task. Baier's account of "meaning" includes reference to the pursuit of worthwhile projects, both in the sense of "projects that affords satisfaction to the pursuer," and in the sense of "morally worthwhile projects"—concerning, for instance, the well-being of others. Questions of the meaning of life, I suggest, are typically questions of how these two sorts of pursuit can be *fused*. I do not think that the use of such words as "fulfilment" and "self-realisation" need force one into a refined form of egoism, nor that the price of avoiding such is to empty the words of all sense. It need not be claimed that the values, duties, etc. involved *derive their force* from their ability to gratify the agent. If a particular agent fails to see the pursuit of moral and social goals as conferring meaning on his life, he need not be lamenting that he has failed to envisage how their pursuit could yield gratification for him. Nor is he necessarily revealing *Akrasia*. He may actually rise to his duty, but he rises to it as one rises from a warm bed to a chill morning's tasks. If he is asking how his duties (and other pursuits) can be represented as of "interest" to him, it need not be in the sense of "interest" that egoism requires, but in the sense *"of concern"* to him. What one often finds, on reading the reflective autobiographies and semi-autobiographies that are undoubtedly our best sourcebooks here, is that in asking questions about the meaning of his life, the author is asking how he can relate the pursuit of various valuable ends to the realising of a certain kind or form of life, the thought of which evokes in him the response: "The pursuit of these goals really concerns me, matters to me!" He may achieve this by way of an imaginative vivifying of the objectives (moral, social, religious . . .) of his policies and pursuits, by summoning up and dwelling upon a vision of the ideal that facilitates self-identification with it. The writer may be helped in this by some extant public myth (compare R. B. Braithwaite on religious belief), or he may elaborate a private myth in which he casts himself in the role of a person dedicated to the pursuit of the valuable ends, whatever they are.[3]

3. Obviously relevant autobiographical quarries include *The Prelude, Dichtung und Wahrheit,* Chesterton's *Autobiography* and Berdyaev's *Dream and Reality.* I have discussed some other aspects of these topics in "Vision and Choice in Morality," *Proc. Arist. Soc. Suppl.* vol. 1956, reprinted in *Christian Ethics and Contemporary Philosophy,* ed. I. T. Ramsey (S.C.M. 1966). The present article is generally complementary to that paper.

I have kept the plural—"goals," "ends"—following the tendency (noted above) to repudiate "monolithic" accounts of the good life. But it needs to be remarked that there are perfectly intelligible occurrences of the singulars in many contexts; occurrences that do not stem from bad metaphysics or bad moral philosophy. A life may be said to acquire "new meaning" through the rallying and ordering of resources that have hitherto been dispersed and conflicting in disunity. This is a sort of "integrating" that is independent of any claims about all goods being ultimately one. Vronsky, in love with Anna Karenina, "felt that all his powers, hitherto dissipated and scattered, were now concentrated and directed with terrible energy toward one blissful aim . . . He knew that . . . all the happiness of life and the only meaning of life for him now was in seeing and hearing her" (I, ch. xxxi). In the same novel, Levin, having seen Kitty passing in a coach, reflects: "in the whole world there was only one being able to unite in itself the universe and the meaning of life for him. It was Kitty" (III, ch. xii).

All these complexities are reflected in the complexity of the criteria we are likely to use in commenting upon or appraising discourse about the meaning a person's life has for him. Where we are moved to disapproval, our criticism may certainly be in part a *moral* criticism: one may disapprove, for instance, of a person's commitment to the Don Juan pattern of life, its goals and priorities. But such criticism may involve other factors besides the moral. Suppose someone does take the Don Juan pattern as fitting his life as lived at present and as setting "tasks" for his future. But suppose also that in this particular case, the life as it is being lived is totally ineffectual in the relevant respects. We shall want to say, "*This* meaning cannot be given to *this* life"; and the cost of trying to impose it is a falsification of the course the life is taking in fact. The effect of such blunders in self-commitment to a pattern of life may be grotesque or pathetic, or, as in the case of the other Don—Don Quixote—it may be richly comic. Or again, although we might not wish to speak of falsification, of events and pattern at loggerheads, we might detect that there is a self-conscious, self-dramatising and only half-sincere playing-out of a role. Whereas, on the other hand, we should count integrity, the refusal of stereotyped, paste-board *personae,* as pre-eminent among grounds for approval in this domain. The agent himself may come to see, in a dawning self-knowledge, that some attempt to give meaning or purpose has been proceeding on the wrong lines. Once more, in *Anna Karenina* (Part II, ch. xxxv), Kitty's attempt to find meaning through adopting Varenka's religious way of life ends unhappily. She decides it has not risen above the level of an imitation, and has involved pretence and self-deception. Some pages later, Anna reflects on her efforts to love Karenin:

"Have I not tried, . . . with all my might, to find a purpose in my life? Have I not tried to love him . . . ? But the time came when I could no longer deceive myself . . ." (Part III, ch. xvi). Where we can estimate them, the ease or difficulty, "slickness" or strenuousness involved in the quest for meaning may figure in criticism. We sometimes say that someone has found meaning in too facile or superficial a way, or that an autobiography discloses a "lack of probing," "a relaxed mood" or a cocksureness that replaces the tension of real self-exploration.[4]

The pursuit of meaning, in the senses we have been examining, is a sophisticated activity, involving a discipline of attention and imagination. Is this true of *all* relevant senses? The case of Tolstoi and the peasants, as discussed by Flew, suggests otherwise. To look back at that may help to plot further ramifications.

What makes it plausible to say that the peasants knew the meaning of life, and in what sense, if any, did they *not* know it? On Flew's Rylean account (*ibid.* p. 116) "the secret of the peasants . . . [is] knowledge of *how* to go on living, [and this is] only another way of saying that they . . . enjoy rude mental health." This "no nonsense" analysis relies on a very weak sense of "knowing how." It is not being claimed that the peasants had mastered techniques for banishing depression—for they did not suffer from it; nor even rules, skills by which to steer their lives round the hazards of depression and mental arrests. They could hardly know of such states of mind as Tolstoi's or J. S. Mill's or Coleridge's in "dejection." They are unaware of the perils they have missed. If for the peasants "knowing the meaning of life" equals "knowing how to live," it amounts to no more than this: that as a matter of fact the peasants were not vulnerable to the malaise that Tolstoi suffered. One might want to say, "This is surely not a 'knowing' *at all*," but just a fortunate combination of circumstances *happening* to produce "rude mental health." It is true, however, that oridinary language allows this weak sense of "knowing how," a sense that is consistent with Flew's account of the peasants. It is applicable, for instance, to the baby who "knows how to cry," even to the bird who "knows how to build a nest."

On the other hand, if one's aim is (like Flew's) "to throw light upon the meaning of the question: 'What is the meaning of life?'" (*ibid.* p. 110), one must keep in mind that a quite important element of meaning is not covered by this weak sense of "knowing how . . . ," "knowing how to live," etc. Tolstoi may well have been mistaken in thinking that the peasants had an esoteric knowledge that he lacked, a knowledge that gave a key to the problem of life's meaning. But, in thinking so, he was

4. Roy Pascal, *Design and Truth in Autobiography* (1960), pp. 181, 191.

testifying to that side of the logic of "the meaning of life" that has to drop out in the case of the peasants: namely, that questions about the meaning of life involve a problem, see life as a problem, and involve a search for an answer. This is so frequently presupposed in discourse about the meaning of life, that such discourse is felt as curiously attenuated, if it is thought away. If we do take the problematic context as built in to the logic, then the sense of "knowing how to live" that is correlative with solving the problem and coming to know that meaning, will be a stronger sense than can be used of the peasants. It will involve awareness of the hurdles, the threats of futility, and the devising of tactics to overcome them.

Could a man's life have or fail to have meaning, without his knowing that it did or did not have meaning? This question could be answered either way, according to the interpretation given to "the meaning of life." We can answer Yes: his own awareness of his life's meaning is not a necessary condition for its being meaningful. This is so, if "having meaning" is equated with "contributing to valuable projects," "achieving useful results." But it may be felt, again, that such an answer goes against the grain of language: that it is too odd to say, for instance, "White did not himself find the meaning of his life: but Black (White's biographer, writing after White's death) did find it." Finding the meaning of White's life may be deemed to be something only White is logically able to do. The giving of meaning to life is seen as essentially a task for the liver of it. Conversely, in the case of people who have not worked at it, seen it as a task, who have been unreflectively happy or unhappy, it would be most natural to say that they have neither found nor failed to find the meaning of life.

It must, of course, be admitted that the peasants had something Tolstoi wanted to have but did not have. In the sense we have been elucidating, if Tolstoi succeeded in getting it, he could be said to have found the meaning of life, but we should not say the same of the peasants, since they had never been conscious of a problem. If this is logically curious, it is not unintelligible. Were life never problematic, were people never subject to arrests of life, it is unlikely that we should ever have acquired the expressions we are discussing. But having once acquired them in the problematic context, they can be extended to other and non-problematic contexts. The former seem to remain primary, however, and it distorts the logic of "the meaning of life" if we take as paradigmatic instances of success in discovering meaning, people who have never been troubled by the problematic aspects of life, its limits, contingency and the like.

(Two smaller, related points may be mentioned in passing. Flew's statement that the characters in Tolstoi's novels know better than Tolstoi is contestable. In the case of Levin there is a good deal about the revelation of "a knowledge unattainable by reasoning." The knowledge of "what we should live for . . . cannot be explained by reason" (Part VIII, chs. xix, xii, etc.), Again, Tolstoi does permit himself to use the expression, "the meaning of life" in a remarkably different sense from that on which our discussion, and Flew's, has centred. When Anna, just before her suicide, discerned the hopelessness of her position, "she saw it clearly in the piercing light which now revealed to her the meaning of life and of human relations" (Part VII, ch. xxx)—a diabolical, not a benign, pattern being revealed. To have seen this "meaning" was not to be enabled to live, but was to judge, on the contrary, that a continuance of life was unendurable. This is an eccentric but noteworthy usage.)

On the highly particularised problems of giving meaning to an individual life, philosophy may not have much to say: but it is certainly concerned with what seem to be general threats to meaningfulness arising out of the human situation as such. For the non-theist, the chief threat may well appear to come from the realisation of mortality. The relation between meaning and mortality is, as we have noted, a focus of attention in current discussions. On the one side are writers (Tolstoi is again among them—as Flew brings out) who in some contexts virtually identify the question of meaningfulness with the question of immortality: deny immortality and you necessarily deny meaningfulness. This account plainly distorts the logic of the question about the meaning of life, not least by reducing its complexity to a single issue of fact. On the other side, a naturalistic account oversimplifies and dogmatises if it claims that "death is irrelevant" (Baier) to questions of value, worthwhileness—and hence to meaningfulness. The argument that values are not devalued by mortality, although unassailable on grounds of logic, may be used to express a naturalism that is more optimistic and brash than it is entitled to be. For a person may heed the argument that mortality and value are compatible, and yet may be burdened with a sense of futility that the argument cannot dispel. His malaise may range from occasional vague misgivings over the worthwhileness of his activities, to a thoroughgoing arrest of life.

We can most usefully consider this sort of malaise at a still more general level. A complaint of meaninglessness can be a complaint about a felt disproportion between preparation and performance; between effort expended and the effect of effort, actual or possible. Yeats expressed the complaint in well-known words: "When I think of all the books I have read, wise words heard, anxieties given to parents, . . . of hopes I have

had, all life weighed in the balance of my own life seems to me *a preparation for something that never happens.*" (Compare again, Tolstoi's question, "What does it *lead to?*")

A useful vocabulary in which to discuss this can be borrowed from
aesthetics. In Monroe Beardsley's book, *Aesthetics* (New York, 1958, pp.
196 f.), he distinguishes, in a musical composition, passages that have
"Introduction quality" and passages that have "Exhibition quality": the
one is felt as leading up to the other, the first as preparation and the second as fulfilment. He quotes, *apropos,* an amusing criticism by Tovey of
Liszt's tone poem, *Ce qu'on Entend sur La Montagne.* The work consists
of "an introduction to an introduction to a connecting link to another introduction . . . ," etc. Beardsley adds: "When [the tone poem] stops it
is still promising something that never arrives."

In music this sort of disappointment can be due to more causes than
one. It may be the listener's expectations, not the music, that are at fault.
He may be simply misinformed about the programme he is to hear; ex-
pecting a symphony, but getting only a concert overture. A musical idiom
may disappoint him: he awaits a fully-fashioned melody in the style of
Brahms, but—since the music happens to be in the style of Webern—it
provides none. In different cases, different kinds of correction are required:
to study the programme, to familiarise himself with musical development
after Brahms. What sounded prefatory can come to sound performatory.
But, as in the Liszt-Tovey case, the fault may lie with the music, and
nothing the listener can do can give it an exhibition quality.

How far can the analogy be applied? To draw first its limits: a piece
of music is "given" to a listener in a way that his life, being partly shaped
by his choices, is not given. The nearest to a given in the latter case is an
awareness of the general conditions and scope of human existence, of what
is it open for men to do and to be, of what they can reasonably count on
doing and being. Suppose a person who has lost belief in immortality has
a sense of futility due to the feeling that life is all prefatory, has no exhibition quality. An analytical philosopher points out to him that things
can be valuable although not eternal, and so on. This is rather like pointing out that a symphony was never on the programme, but that other valuable things were. The implication is that just as a satisfactory short piece
of music can be heard as having exhibition quality, so can life within the
limits of earthly existence be perceived as having it.

This suggestion may dispel the sense of futility, or it may not. It may
dispel it, because to some considerable extent it does lie within our power
to take a particular experience either in a prefatory or exhibitory way, although how this is done we cannot inquire here in any detail. In general,
and very obviously, many activities can be understood both as leading to

anticipated worthwhile ends, and as worthwhile in themselves also. A person may sometimes successfully counsel himself to dwell less upon the possible future effects of his various activities, and relish the present activities themselves. (To do so, there must, admittedly, be something relishable about the activities.)

Suppose, then, the suggestion does not dispel the sense of futility. To follow the analogy a step further: life may be found recalcitrant in the way that Liszt's tone-poem was recalcitrant to Tovey. Clearly there are experiences, of prolonged and unrelievable ill-health, for instance, that can be regarded as valuable, in the sense of disciplinary, only so long as death is not believed to be the end. Lacking such belief, a person cannot simply be counselled to endow a life pervaded by pain with exhibition quality. It is of such suffering that people tend to use words like "pointless" and "senseless." If this experience is followed by no "exhibition-section," it is unredeemably futile.

What follows? The existence of unsuccessful musical compositions exposed to the Tovey criticism does not devalue those that are not exposed. Must one say any more, in the case of lives, than that some lives are without meaning, in the present sense; but that others are meaningful? Or, in terms of the earlier discussion, that some people are confronted with a stiffer problem than others in giving meaning to their lives, and that the task may sometimes be impossible for them. This would be enough to qualify an over-optimistic naturalism that implies that (where meaning is concerned) everything remains as it was in the days of belief in immortality and Providence.[5] As an index of the difficulty of nicely proportioning prefatory and exhibitory quality in a life, one may instance an aesthetic theory like John Dewey's, according to which the distinctiveness of art is its capacity to provide just such completeness and balance in experience—in contrast to the ordinary experience of life, which seldom can maximise these qualities and can never *guarantee* them.[6]

If it is a condition for life-in-general to be meaningful that these qualities *should* be guaranteed, then life-in-general is without meaning to the naturalist or agnostic—though individual lives may happen to attain it, by practical wisdom and good fortune.

Judgments about the value or futility of human projects are problematic in yet another way. Our appraisal of the value of our activities tends

5. In the studies from which this discussion started, Baier and Nielsen, e.g., acknowledge that, on their analysis, particular lives may, in particular extreme situations, be "meaningless."
6. For instance: "Art celebrates with peculiar intensity the moments in which the past reënforces the present and in which the future is a quickening of what now is" (Dewey, *Art as Experience* [1934], p. 18).

to vary according to the "backcloth" against which they are viewed. To the naturalist, human endeavour viewed *sub specie aeternitatis,* may seem to shrivel, frighteningly. (Compare Sartre's troubled musings on the heat-death of the sun, in his autobiographical essay, *Les Mots,* p. 208.) He may nonetheless judge as entirely worthwhile some social reform, viewed against the backcloth of a dozen years, or months, of social abuse. Yet it is not only a vestigial romanticism that prompts one to give privileged importance to the widest and broadest backcloth. For the more fully synoptic one's view, the more confident one becomes that one is reckoning with all possible threats of vilification. The movement is towards the discarding of blinkers and frames that artificially confine attention to a narrow context and which might equally artificially boost a sense of importance and worthwhileness. Only: on naturalist assumptions, the *sub specie aeternitatis* view must at least seem to vilify, by revealing human history in an ocean of emptiness before and after. Some of the logical darkness around the meaning-of-life questions comes from the uneasy awareness that there *are* alternative views, perspectives, more or less synoptic or selective in different ways, giving very different answers to questions of value, importance, futility. From this bewildering diversity the questioner seeks some release.

He may come to the problem with the belief or half-belief that there must be an authoritative view; and that if life is to be shown to be meaningful, this authoritative view will also be a value-confirming, value-enhancing view. On authoritativeness, however, I think he is asking for what cannot be given. Either a "view" is a sheer psychological fact about how someone sees human activity, in which case it has nothing that can properly be called "authority"—though it may have strikingness and imaginative force. Or, if we judge that a particular view does have authority, this is to make a judgment of value and not simply to describe one's imaginings. Now, the estimate of the importance of human life implied in the "authoritative" view may or may not conflict with the estimate one makes on other bases. If it does not conflict, there is no further problem: if it does conflict, then we must ask, Which has to yield, the judgment that the perspective is authoritative, or the independent (perhaps workaday) estimate of worthwhileness? The criteria for authoritativeness, crucially, are no less challengeable than the independent value-judgment: only by the agent weighing up, arbitrating between, or oscillating between, the two clashing sets of evaluations, can he deal with his dilemma. If he resolves the dilemma, it is resolved by his own autonomous value-judgments: and therefore the promise of having it resolved for him by some self-evident authoritativeness is not fulfilled.

Despite this argument, the naturalist ought to admit a substantial difference between his position and that of the Christian theist. There must in fact remain, with the naturalist, an uncomfortable tension or conflict between the "close-up," anthropocentric view or perspective that can sustain his sense of meaningfulness and worthwhileness, and on the other hand his sense of intellectual obligation to the objective, scientific and anti-anthropocentric view—which tends to vilify, if not logically, then psychologically. The Christian is not exposed to this tension in the same way or to the same extent. The doctrines of divine creation and of incarnation combine to rule out the judgment that, in leaving the arena of the human, one is leaving simultaneously the theatre of mind and purpose and value. For the theist, that is, there is the implicit promise of a *harmony* of perspectives. To challenge that is to challenge the ultimate coherence of theism.

Once again, the meaning-of-life vocabulary may be so used (with exacting criteria of application), that only where such a harmony is promised can human life be properly called meaningful. Or at least the reluctance of a theist to concede that life can be meaningful to the unbeliever may reflect this, perhaps unanalysed but sensed, difference between their views of the world.

In the greater part of this discussion we have been assuming that the paradigm case of a satisfactory answer to problems about the meaning of life is contained in Christian theism, and that the question for the non-Christian is whether anything at all can be rescued from a collapse of meaning. This assumption, however, may be questioned in various ways. Two of these must be briefly mentioned.

That Christianity is able to provide a satisfactory answer is implicitly denied by those critics who claim that no hereafter, even of endless beatitude, could compensate for some of the evils actually endured by men here and now.[7] To revert to the musical analogy: in a "meaningful" work, the exhibition passages fulfil and complete the anticipatory passages, which are unsatisfying by themselves. We can conceive of an anticipatory passage, however, which is so atrocious musically that nothing that followed it, even of high quality, could be said to fulfil or complete it. It may be argued that some "passages" in some people's lives are so evil that nothing could conceivably justify them. As with a *privatio* theory of evil, so with analysis in terms of "introductory quality": it makes out evil to be what we cannot by any means always see it to be, as able to be supplemented so as to become an ingredient in an aesthetically and morally valu-

7. Dostoevsky's is the classic statement: see *The Brothers Karamazov*, Book V, chap. 4.

able whole. Here, therefore, the meaning-of-life problem runs into the problem of evil. To the extent that one is baffled by the problem of evil, to the same extent one must be baffled by the meaning of life. We cannot open up that problem here.

Secondly, in his aggressively anti-theistic argument, Kurt Baier claims that theism seeks to give meaning and purpose to human life in a morally objectionable way. Understanding one's life as given purpose by God involves a thoroughgoing self-abasement or self-annulment quite incompatible with the stance of a moral agent. This self-abasement is a correlative of the recognition of God's holiness (on current accounts), the realisation in numinous dread that God is "wholly other."

The objection is impressive and serious, and certainly effective against some forms and formulations of theism. If the "otherness" of God and the *tremendum* side of the mystery are one-sidedly stressed, then human moral judgment and divine purpose are bound to be seen as incommensurable. An attitude of worshipfulness will involve an abrogating of moral autonomy.

It seems equally clear, however, that this is not the only posture a theist can adopt. Worship is by no means a simple, single-stranded concept. To worship is not only to yield to an overwhelming of the intellect or engage in unthinking, undiscerning adulation. Something very different from this is expressed, for instance, in the *B Minor Mass* or Stravinsky's *Symphony of Psalms*. An act of worship need not be a submitting of oneself, without insight, to the wholly inscrutable. Moral perfection and beauty, fused in an intensifying strangeness, are being celebrated, as these are believed to inhere in God. Evaluative reflection is being exercised: it is very far from abdicating. The worshipper is not passively, heteronomously, accepting claims about divine greatness: he is actively and autonomously recognising and relishing them. His worship is not a preface to a life of submission to commands *ab extra*. The very ingredients of his worship itself—in attitude and feeling—become part of his own inner life, the stillness of soul, the wonderment and solemnity.

The element of strangeness, of *mysterium*, has however a further role to play. Without cancelling the values, moral and aesthetic, that are celebrated in the act of worship, it powerfully intimates that they are nonetheless open to further transformation. It forbids thinking that the human vision of the good is precisely congruent with the nature of divine goodness. There is both recognition of affinity and awareness of disparity. But the latter need not be taken as a command to surrender autonomy: rather to extend progressively the zone of "recognition."

If it is true that a worshipful stance does not involve abrogating moral autonomy, neither need an acceptance of a divine purpose or meaning to the worshipper's life. This would indeed involve a regulating of one's life, a limiting of one's projects to what, rightly or wrongly, one judged to be compatible with God's will and intention. But such a limiting or circumscribing is a common feature of familiar moral situations, where one acts say, as a member of a trust, or has a special status in a group, a status that prescribes special obligations. In many, if not all such cases— and the case of divine purpose is no exception—the regulating of action is necessary, if some good or goods is to be realised. A rational person will make it his business to understand how the relevant moral institution does bring about that good. In the theistic case one needs reasons for believing that God's wisdom, goodness and power are such that the following of his will is a surer way to achieving goods in general than by carving out patterns of life that ignore or transgress his will.

Nevertheless, this sort of language quickly brings one to difficulties more stubborn than the original problem of autonomy. The fact that some worshipful activity expresses morally and aesthetically valuable states of mind does nothing to guarantee the existence of the God who is the primary object of the worship. It cannot show that God, if he exists, has the qualities celebrated in the act of worship, that these qualities are mutually compatible in a single being or that God can have these qualities and the world be the sort of world it is. And we confront the problem of how one can meaningfully speak of God, an infinite being, as having such features of finite person-hood as willing, intending, purposing. In a word, there remain difficulties enough in the theistic conception of the meaning and purpose of life, but they are the general difficulties of theism itself. The objection that the theist must abdicate his moral status is by no means decisive against all forms of theism, although valid against some.[8]

At the start of this article I mentioned one terse and gruff dismissal of the whole topic, the meaning of life: "Life is not a statement, is not a linguistic entity, so cannot strictly be said to have or lack meaning." We evaded this objection by way of a part-equating of "meaning," in this context, with "point" or "purpose." Yet it is worth asking whether, in discussing the meaning of life, some analogy with linguistic meaning may not often be operative—either in the background or the foreground of attention, helpfully or misleadingly. In what ways could it operate?

The words in a meaningful sentence "cohere." Words in a random list do not: like the events of a "meaningless" life, they merely succeed each other discretely and atomistically, and are no more. The "past" of a

8. Further arguments on this topic can be found in C. C. J. Webb's criticism of N. Hartmann, *Religion and Theism* (1934), pp. 69–86, 87, 144 to end.

piece of discourse to which one is listening is felt as active and as bearing upon the presently uttered words. It is not a "lost" past. Neither is it an obsessively dominant past, that imposes a static and completed pattern in place of the developing and novel pattern that the piece of discourse builds up. Certainly life is not a piece of discourse, and the motifs that supply coherence and continuity to it are very different from a rule-governed syntax. But if not pressed to breaking-point, the analogy may have a regulative function in the managing of our relation to our own past. Extreme cases of *failure* in this task can be seen in schizophrenics to whom every day is "a separate island with no past or future," and who lack all sense of continuity, coherence; and, at the other pole, those who feel "nailed" to their past, unable to distance it at all.[9]

The vocabulary of "meaning," however, can obviously do no detailed regulating or managing by itself. If it appears to do much more than hold the bare form of the task before the attention, it may be spellbinding its user into illusion. It may suggest that the pattern of our life as a whole has a unity, and an availability to recollection *in* its unity, beyond what is in fact possible. We may say of apparent unity and coherence in life what Proust said of "the total value of our spiritual nature": "At whatever moment we estimate it, . . . it is more or less fictitious; . . . notwithstanding the long inventory of its treasures, . . . now one, now another of these is unrealisable . . . It is an illusion that our inward wealth is perpetually in our possession."[10]

The linguistic analogy may lure one on still further. The words of a sentence are uttered as the clock ticks, and the events of a life occur as the clock ticks: but the meaning of the sentence is not an episode in time—and the meaning of the life . . . ? To yield to the suggestion, and to say "timeless" of the patterns of continuities, overlapping motifs of a life would be no more illegitimate than to say the same thing of logical relations and patterns. Nevertheless, one cannot use the language of "timelessness," in respect of life, without evoking a whole syndrome of distinctive aspirations, for "deliverance from time," for experience as a *"totum simul,"* for "eternal life": and, seductive though these are, would they not surely be out of place in such reflections?

Yet some writers might not hurry to exorcise them. Santayana, for instance, argued that eternal life cannot be satisfactorily conceived as an endless succession of events, nor, coherently, as a *totum simul*; and concluded that eternity is experienced only in the contemplating of timeless

9. On these forms of schizophrenia, see R. May, *Existence* (1958), pp. 66 (quoting Minkowski), 68.

10. *Sodome et Gomorrhe,* vol. 1, p. 213 (Eng. tr., *Cities of the Plain,* vol. 1, p. 218).

objects and structures.[11] To contemplate the web of one's life's aims, its themes and articulating images, might count, on such a view, as affording as much fulfilment of the longing for eternity as the nature of things affords, everything beyond that being ultimately delusory.

It is natural and not at all absurd for a naturalistic philosopher to seek some (inevitably limited and provisional) substitute for a metaphysical or religious doctrine of eternal life. If he judges that doctrine to be logically incoherent anyway, there is all the less reason to protest if he chooses to savour those facets of experience that give a *partial* backing to the doctrines. The important thing is that he should himself be under no illusions. When he uses the vocabulary of timelessness, the "metaphysical pathos" of his discourse must be appropriate to his real beliefs and must not borrow illegitimate splendour from the theism and mysticism he rejects.

It is possible (and we have remarked on this briefly already) to take note of all the senses we have been discussing, senses in which life may be held to have meaning within a non-theistic philosophy, to share, at least in outline, the value-theory of the philosophers with whom we started, and yet to insist that life and individual lives are without meaning—"really." What is a person doing who insists on this answer?

He may be claiming that life could be thought of as having meaning only so long as that meaning was believed to be a matter for discovery, not for creation and value-decision. He may be claiming that, to be meaningful, life would have to be *comprehensively* meaningful and its meaning invulnerable to assault. Worthwhile objectives must be ultimately realisable, despite appearances. If the Kantian postulates cannot be made, meaningfulness goes: it is for him an all-or-nothing matter. Otherwise expressed, the question about meaning can be put as a question about contingency and fortuitousness. If our attempts to impose or discover coherence and purpose in a life can operate only within the narrow *Spielraum* left by an unplanned and patternless "facticity," in the end by accidents of birth and death, then the language of meaning is far more misleading (sardonically so) than it is useful.[12] If Nietzsche speaks of how desirable it is to "die at the right time," of the "free death" that "consummates," he succeeds only in throwing into stronger relief the impossibility of guaranteeing anything of the sort, and we recall his own lingering ("meaningless") end as a case in point.

11. *The Philosophy of Santayana*, ed. I. Edman (1953), pp. 206–11.
12. Sartre: "All existing things are born for no reason, continue through weakness and die by accident. . . . It is meaningless that we are born; it is meaningless that we die."

The vocabulary of meaning can thus be rejected *en bloc,* as for instance atheist existentialists have tended to reject it. In criticism, perhaps only this can be said: that it is a pity to expend in one rhetorical gesture a piece of discourse that can be used to express important distinctions between and within individual human lives. The gesture, though impressive, is extravagant and linguistically wasteful. But, so far as the rejection is a piece of (negative) verbal stipulation, one is free to opt for it: the language of the meaning of life is not indispensable. Alternatively, as we saw, one may prune and rationalise and redefine the vocabulary, restricting it to judgments about purposefulness or knowledge how to live. Perhaps there is most to be said for the third option, an elucidatory and conservative tracing of senses of "the meaning of life" in literature and philosophy, and an attempt to see how differently the tasks of a life can look when they are viewed in the light of the many analogies of meaning.

Our sample of these analogies in this article, if not random, is certainly not complete.[13]

13. A version of this study was included in my Stanton Lectures, at the University of Cambridge, Lent Term 1966.

≫ 16 ≪

PAUL EDWARDS

Why?

Lack of clarity about the uses of the word "why" is responsible for confusion on a number of philosophical fronts. In this article we shall confine ourselves to two groups of topics where greater attention to the proper and improper behavior of this word might well have avoided the adoption of misguided theories. There is, first, the contrast, or the alleged contrast, between the "how" and the "why" and the view, shared by writers of very different backgrounds, that science can deal only with how-questions. Second, there are certain "ultimate" or "cosmic" questions, such as "Why do we exist?" or, more radically, "Why does the world exist?" or "Why is there something rather than nothing?" Some, like Schopenhauer and Julian Huxley, regard these questions as unanswerable; others, like Gilson and Copleston, believe that they can be answered; but whether these questions can be answered or not, it seems to be widely agreed that they are very "deep." These questions, in the words of the British astrophysicist A. C. B. Lovell, raise problems "which can tear the individual's mind asunder" (*The Individual and the Universe,* New York, 1961, p. 125). Speaking of the question "Why is there something rather than nothing?" Heideger first remarks that it is "the fundamental question of metaphysics" and later adds that "with this question philosophy began and with this question it will end, provided that it ends in greatness and not in an impotent decline" (*An Introduction to Metaphysics,* p. 20).

HOW AND WHY

The contrast between the how and the why has been insisted on for two rather different reasons. Some writers have done so in the interest of religion or metaphysics. Their position seems to be that while science and empirical research generally are competent to deal with how-questions, the very different and much deeper why-questions are properly the concern of religion or metaphysics or both. Thus, in a widely read book the British psychiatrist David Stafford-Clark insists that the confusion between the how and the why is the "fundamental fallacy" behind "the whole idea that science and religion are really in conflict at all" (*Psychiatry Today*, Harmondsworth, England, 1952, p. 282). Freud in particular is accused of commiting this fallacy in his antireligious writings. Stafford-Clark is not at all opposed to Freudian theory so long as it confines itself to the how of psychological phenomena. Psychoanalysis cannot, however, "begin by itself to answer a single question as to why man is so constructed that they should happen in this way" (*ibid.*, p. 287). Although he repeatedly expresses his own fervent belief in God, Stafford-Clark unfortunately does not tell us how religion answers the question why man is "constructed" the way he is. Perhaps he would answer it along the lines in which Newton answered a similar question about the sun. "Why is there one body in our system qualified to give light and heat to all the rest," Newton wrote in his first letter to Richard Bentley, "I know no reason, but because the author of the system thought it convenient" (*Opera*, London, 1779–1785, Vol. IV, pp. 429 ff.).

Similar views are found in the writings of many professional philosophers. Thus, writing of Newton's work on gravitation, Whitehead observes that "he [Newton] made a magnificent beginning by isolating the stresses indicated by his law of gravitation." But Newton "left no hint, why in the nature of things there should be any stresses at all" (*Modes of Thought*, New York and Cambridge, 1938, pp. 183–184). Similarly, discussing the limitations of science, Gilson declares that "scientists never ask themselves *why* things happen, but *how* they happen. . . . Why anything at all is, or exists, science knows not, precisely because it cannot even ask the question" (*God and Philosophy*, New Haven, 1959, p. 140). For Gilson the two topics mentioned at the beginning of this article appear to merge into one. The why of particular phenomena, he seems to argue, cannot be determined unless we answer the question "why this world, taken together with its laws . . . is or exists" (*ibid.*, p. 72).

Among those who have asserted that science can only deal with how-questions there are some who are not at all friendly to metaphysics or religion. These writers usually add to their remarks that science cannot handle why-questions the comment that no other enterprise fares any better. This "agnostic positivism," as we may call it, goes at least as far back as Hume. We know, he writes, that milk and bread are proper nourishment for men and not for lions or tigers, but we cannot "give the ultimate reason why" this should be so (*An Inquiry Concerning Human Understanding*, Sec. IV,

Part I). Hume seems to imply that this unhappy state can never be remedied, regardless of the advances of physiology or any other science. Several writers in the second half of the nineteenth century advanced this position under the slogan "The task of science is to describe phenomena, not to explain them." Ernst Mach, Gustav Kirchhoff, and Joseph Petzoldt were among the best-known figures in central Europe who advocated this view. In England, Karl Pearson, its most influential exponent, conceded that there was no harm in speaking of "scientific explanations" so long as "explanation" is used "in the sense of the descriptive-*how*" (*The Grammar of Science,* Everyman edition, 1937, p. 97). We can indeed "describe how a stone falls to the earth, but not why it does" (*ibid.,* p. 103). "No one knows why two ultimate particles influence each other's motion. Even if gravitation be analyzed and described by the motion of some simpler particle or ether-element, the whole will still be a description, and not an explanation, of motion. Science would still have to content itself with recording the *how.*" No matter how far physics may progress, the why will "remain a mystery" (*ibid.,* p. 105).

It is important to disentangle purely verbal from substantive issues in all of this. Insofar as the various writers we have quoted merely wish to assert that causal statements and scientific laws in general are contingent and not logically necessary propositions, little exception could be taken to their remarks. However, they are, or at least they appear to be, saying a great deal more. They all seem to agree that there is a class of meaningful questions, naturally and properly introduced by the word "why" in one of its senses, which cannot be answered by the use of empirical methods. Writers belonging to the first group claim that the answers can be obtained elsewhere. The agnostic positivists maintain that human beings cannot obtain the answers at all.

It is this substantive issue which we shall discuss here, and it is necessary to point out that there are numerous confusions in all views of this kind. To begin with, although this is the least important observation, "how" and "why" do not always have contrasting functions but are in certain situations used to ask the very same questions. Thus, when we know or believe that a phenomenon, A, is the cause of another phenomenon, X, but at the same time are ignorant of the "mechanics" of A's causation of X, we indifferently use "how" and "why." We know, for example, that certain drugs cure certain diseases, but our knowledge is in a medical sense "purely empirical." Here we would be equally prepared to say that we do not know "why" the drug produces the cure and that we do not know "how" it does this. Or, to take a somewhat different case, it is widely believed that cigarette smoking is causally connected with lung cancer. It is also known that sometimes two people smoke the same amount and yet one of them develops lung cancer while the other one does not. In such a case the question naturally arises why cigarette smoking, if it is indeed the cause at all, leads to cancer in one case but not in the other. And we would be just as ready to express our ignorance or puzzlement by saying that we do not know how it is as by saying that we do not know why it is that smoking produced cancer in the first man but not in the second. In all such cases it is clear that science *is* in principle competent to deal with the "why" no less than with the "how," if only because they are used to ask the very same questions.

It is undeniable, however, that in certain contexts "how" and "why" are used to ask different questions. This contrast is most obvious when we deal with intentional, or more generally with "meaningful," human actions. What seems far from obvious, what in fact seems plainly false, is that empirical methods are not in principle adequate to determine the answers to why-questions in these contexts. Let us take as our example the recent theft of the Star of India sapphire and other gems from the Museum of Natural History in New York. We can here certainly distinguish the question why the burglary was committed from the question how it was carried out. The latter question would concern itself with the details of the act—how the thieves got into the building, how they immobilized the alarm system, how they avoided the guards, and so on. The why-question, by contrast, would inquire into the aim or purpose of the theft—were the thieves just out to make a vast amount of money, or were there perhaps some other aims involved, such as proving to rival gangs how skillful they were or showing the incompetence of the police force? Now, the aim or purpose of a human being is surely not in principle undiscoverable, and frequently we know quite well what it is. The person himself usually, though not always, simply knows what his aim is. An orator, for example, who is advocating a certain policy, ostensibly because it is "for the good of the country," may at the same time know perfectly well that his real aim is personal advancement. It used to be said that in such situations a human being knows his own purpose by means of "introspection," where introspection was conceived of as a kind of "inner sense." This way of talking is not inappropriate to situations in which somebody is confused about his own motives, for then special attention to his own feelings, resembling in some ways the effort to discriminate the detailed features of a physical scene, may well be necessary in order to ascertain his "true" aims. Much more commonly, however, a human being simply knows what his aims are, and it would be much better to say that he knows this "without observation" than that he knows it by introspection. In order to find out the purpose of somebody else's action, it is in countless instances sufficient to ask the person a direct question about his aim. Where the agent's veracity is suspect or where a person is the victim of self-deception, it is necessary to resort to more elaborate investigations. In the former type of case one might ask the agent all kinds of other questions (that is, questions not directly about the purpose of his action), one might interview his friends and acquaintances and other witnesses of his conduct, one might tap his telephone and employ assorted bugging devices, and one might perhaps go so far as to question him after the administration of "truth" drugs. In the latter type of case it may not be possible to ascertain the real purpose unless the person undertakes psychiatric treatment. While the practical difficulties in the way of discovering the purpose of an action are no doubt insurmountable in many cases of both these types, empirical procedures are clearly in principle adequate to this task.

We also contrast how- and why-questions when the latter are not inquiries into the purpose of any agent. Here, however, "how" has a different meaning from any previously discussed. In all examples so far considered, how-questions were in one way or another *causal* questions—"How did the thieves carry out their plan of stealing the Star of India?" is a question about the means of achieving a certain goal, and "How is it that smoking produces cancer in one man but not in another?" although not a question about means, is nevertheless about the processes leading to a certain result. These causal "hows" should be distinguished from what one may call the "how" of "state" or "condition." "How cold does it get in New York in the winter?" "How does the decline in his powers manifest itself?" "How is his pain now—is it any better?" are examples of the "how" of state or condition, and it is how-questions of this kind which we contrast with nonteleological why-questions—"Why does it get so cold in New York in the winter?" "Why did his powers decline so early in life?" "Why is his pain not subsiding?"

It is sometimes maintained or implied, as in the remarks of Stafford-Clark quoted earlier, that why-questions are invariably inquiries about somebody's purpose or end—if not the purpose of a human being, then perhaps that of some supernatural intelligence. This is clearly not the case. There can be no doubt that "why" is often employed simply to ask questions about the cause of a phenomenon. Thus the question "Why are the winters in New York so much colder than in Genoa, although the two places are on the same geographical latitude?" would naturally be understood as a request for information about the cause of this climatic difference, and it is not necessary for the questioner to suppose that there is some kind of plan or purpose behind the climatic difference in order to be using the word "why" properly. In saying this, one is not begging any questions against the theory that natural phenomena like the cold of the winter in New York are the work of a supernatural being: one is merely calling attention to what is and what is not implied in the ordinary employment of "why" in these contexts.

Let us briefly summarize the results obtained so far: in some situations "how" and "why" are naturally employed to ask the very same questions: when we deal with intentional human actions, we naturally use "why" to inquire about the purpose or goal of the agent and "how" to learn about the means used to achieve that goal; finally, how-questions are frequently used to inquire about the state or condition of somebody or something, while why-questions inquire about the cause of that state or condition without necessarily implying that any purpose or plans are involved. In all these cases it appears to be in principle possible to answer why-questions no less than how-questions, and this without the aid of religion or metaphysics.

THE THEOLOGICAL "WHY"

Let us turn now to what we earlier called "cosmic" why-questions. Two such cosmic "whys" need to be distinguished, the first of which, for rather obvious reasons, will be referred to as the theological "why." Here the questioner would be satisfied with a theological answer if he found such an answer convincing in its own right. He may or may not accept it as true, but he would not regard it as irrelevant.

Gilson, whose remarks on the limitations of science were quoted earlier, immediately supplies the answer to the "supreme question" which science "cannot even ask." Why anything at all exists must be answered by saying:

> [Each] and every particular existential energy, and each and every particular existing thing depends for its existence upon a pure Act of existence. In order to be the ultimate answer to all existential problems, this supreme cause has to be absolute existence. Being absolute, such a cause is self-sufficient; if it creates, its creative act must be free. Since it creates not only being but order, it must be something which at least eminently contains the only principle of order known to us in experience, namely, thought. (*God and Philosophy*, p. 140)

There is no doubt that many people who ask such questions as "Why does the universe exist?" or "Why are we here?" would also, at least in certain moods, be satisfied with a theological answer, though they would not necessarily accept all the details of Gilson's Thomistic theology. It should be emphasized that one does not have to be a believer in God to be using "why" in this way. The American playwright Edward Albee, for example, recently remarked, "Why we are here is an impenetrable question." Everyone in the world, he went on, "hopes there is a God," and he later added, "I am neither pro-God nor anti-God" (New York *Times*, January 21, 1965). Albee's question "Why are we here?" evidently amounts to asking whether there is a God and, if so, what divine purposes human beings are supposed to serve. He does not definitely accept the theological answer, presumably because he feels unsure of its truth, but he does regard it as very much to the point.

It should be observed in passing that people frequently use the word "why" to express a kind of cosmic complaint or bewilderment. In such cases they are not really asking for an answer, theological or otherwise. This use of "why" is in some respects similar to the theological "why" and may not inappropriately be referred to as the quasi-theological "why." A person who is and regards himself as a decent human being, but who is suffering a great deal, might easily exclaim "Why do I have to suffer so much, when so many scoundrels in the world, who never worked half as hard as I, are having such a lot of fun?" Such a question may well be asked by an unbeliever who is presumably expressing his regret that the workings of the universe are not in harmony with the moral demands of human beings. Even when believers ask questions of this kind, it may be doubted that they are invariably requesting information about the detailed workings of the Divine Mind. In the deeply moving first-act monologue of *Der Rosenkavalier*, the Marschallin reflects on the inevitability of aging and death:

> I well remember a girl
> Who came fresh from the convent to be
>   forced into holy matrimony.
> Where is she now?

> How can it really be,
> That I was once the little Resi
> And that I will one day become the old
>   woman?

How, she exclaims, can something like this be? She is far from doubting the existence of God and proceeds to ask:

> Why does the dear Lord do it?

And worse, if he has to do it in this way:

> Why does He let me watch it happen
> With such clear senses? Why doesn't He
>   hide it from me?

The Marschallin obviously does not expect an answer to this question, not, or not merely, because she thinks that the world's metaphysicians and theologians are not quite up to it. She is not, strictly speaking, asking a question but expressing her regret and her feeling of complete helplessness.

However, let us return from the quasi-theological to the theological "why." The difficulties besetting an answer like Gilson's are notorious and need not be reviewed here at length. There are the difficulties, much stressed by recent writers, of saying anything intelligible about a disembodied mind, finite or infinite, and there are further difficulties of talking meaningfully about the creation of the universe. There are the rather different difficulties connected not with the intelligibility of the theological assertions but with the reasoning used to justify them. Schopenhauer referred to all such attempts to reach a final resting place in the series of causes as treating the causal principle like a "hired cab" which one dismisses when one has reached one's destination. Bertrand Russell objects that such writers work with an obscure and objectionable notion of explanation: to explain something, we are not at all required to introduce a "self-sufficient" entity, whatever that may be. Writing specifically in reply to Gilson, Nagel insists that it is perfectly legitimate to inquire into the reasons for the existence of the alleged absolute Being, the pure Act of existence. Those who reject such a question as illegitimate, he writes, are "dogmatically cutting short a discussion when the intellectual current runs against them" (*Sovereign Reason*, Glencoe, Ill., 1954, p. 30). Without wishing to minimize these difficulties, it is important to insist that there is a sense in which the theological why-questions *are* intelligible. The question can be answered for such a person if it can be shown that there is a God. If not, it cannot be answered. Albee and Gilson, for example, do not agree about the truth, or at any rate the logical standing, of the theological assertion, but they agree that it is relevant to *their* cosmic why-question. There is thus a sense in which the questioner here knows what he is looking for.

THE SUPER-ULTIMATE "WHY"

The theological "why" must be distinguished from what we are here going to call the super-ultimate "why." A person who is using "why" in the latter way would regard the theological answer as quite unsatisfactory, not (or not just) because it is meaningless or false but because it does not answer *his* question. It does not go far enough. For granting that there is a God and that human beings were created by God to serve certain of his purposes, our questioner would now ask "Why is there a God of this kind with these purposes and not another God with other purposes?" or, more radically, he would ask "Why was there at some time God rather than nothing?" The Biblical statement "In the beginning God created heaven and earth," Heidegger explicitly remarks, "is not an answer to . . . and cannot even be brought into relation with our question." The believer who stops with God is not pushing his questioning "to the very end" (*An Introduction to Metaphysics*, pp. 6–7). (It is not certain how somebody pressing the super-ultimate why-question would react to the rejoinder of those theologians who maintain that God exists necessarily and that hence the question "Why was there at some time God rather than nothing?" is illegitimate. In all likelihood he would support the view, accepted by the majority of Western philosophers since Hume and Kant, that it makes no sense to talk about anything, natural or supernatural, as existing necessarily.)

There are times when most people would regard these super-ultimate why-questions as just absurd. Stafford-Clark himself speaks with impatience of the "rumination" and the tedious and interminable speculations of obsessional patients. " 'Why is the world?' was a question to which one patient could find no answer but from which he could find no relief" (*Psychiatry Today*, p. 112). Yet, at other times, most of us are ready to treat these why-questions as supremely profound, as riddles to which it would be wonderful to have the answer but which, because of our finite intellects, must forever remain unsolved. It is true that certain philosophers, like Schelling and Heidegger, who have frequently been denounced as obscurantists, have laid special emphasis on super-ultimate why-questions: but it would be a total misunderstanding of the situation to suppose that more empirical philosophers, or indeed ordinary people, are not given to asking them or to treating them with great seriousness. It is almost unavoidable that any reasonably intelligent and reflective person who starts wondering about the origin of the human race, or animal life, or the solar system, or our galaxy and other galaxies, or about the lack of justice in the world, the brevity of life, and seeming absolute finality of death, should sooner or later ask "Why this world and not another—why any world?" The scientist Julian Huxley is as far removed in temperament and philosophy from Heidegger as anybody could be. Yet he also speaks of the "basic and universal mystery—the mystery of existence in general . . . why does the world exist?" For Huxley it is science which "confronts us" with this mystery, but science cannot remove it. The only comment we can make is that "we do not know." We must accept the existence of the universe "and our own existence as the one basic mystery" (*Essays of a Humanist*, London, 1964, pp. 107–108). Ludwig Büchner was a materialist and an atheist, and yet he repeatedly spoke of the "inexplicability of the last ground of things." Nor are super-ultimate why-questions confined to those who do not believe in God or who have no metaphysical system. Schopenhauer was supremely confident that his was the true metaphysic, but he nevertheless remarks in the concluding chapter of his main work that his "philosophy does not pretend to explain the existence of the world in its ultimate grounds. . . . After all my explanations," he adds, "one may still ask, for example, whence has sprung this will, the manifestation of which is the world. . . . A perfect understanding of the existence, nature, and orgin of the world, extending to its ultimate ground and satisfying all demands, is impossible. So much as to the limits of my philosophy, and indeed of all philosophy" (*The World As Will and Idea*, 3 vols., translated by R. B. Haldane and J. Kemp, London, 1883, Ch. 50). Similarly, Voltaire, who was a firm and sincere believer in God and who never tired of denouncing atheists as blind and foolish, nevertheless asked, at the end of the article "Why?" in his *Philosophical Dictionary*, "Why is there anything?" without for a moment suggesting that an appeal to God's creation would be a solution. William James, too, although he repeatedly defended supernaturalism, never claimed that it provided an answer to the question "How comes the world to be here at all instead of the nonentity which might be imagined in its place?" Philosophy, in James's opinion, whether it be naturalistic or supernaturalistic, "brings no reasoned solution" to this question, "for from nothing to being there is no logical bridge" (*Some Problems of Philosophy*, New York, 1911, pp. 38–40). "The question of being," he observes later in the same discussion, is "the darkest in all philosophy. All of us are beggars here, and no school can speak disdainfully of another or give itself superior airs" (*ibid.*, p. 46).

Having pointed out how widespread is this tendency to ask and take seriously the super-ultimate why-question, it is necessary to explain why, in the opinion of a number of contemporary philosophers, it must nevertheless be condemned as meaningless. It is the mark of a meaningful question, it would be urged, that not all answers can be ruled out *a priori*; but because of the way in which the super-ultimate why-question has been set up, it is *logically* impossible to obtain an answer. It is quite clear that the questioner will automatically reject any proposed answer as "not going back far enough"—as not answering *his* why. "All explanation," in the words of Peter Koestenbaum, an American disciple and expositor of Heidegger, "occurs within that which is to be explained . . . so the question applies to any possible answer as well" ("The Sense of Subjectivity," p. 54), that is, there *cannot* be an answer. If, however, a question can be put at all, to quote Wittgenstein,

> then it *can* also be answered . . . doubt can only exist where there is a question; a question only where there is an answer, and this only where something *can* be *said*. (*Tractatus Logico-philosophicus*, 6.5 and 6.51)

It must be emphasized that the super-ultimate "why" does *not* express ignorance about the "early" history of the universe. Büchner, for example, had no doubt that matter was eternal and that nothing which could be called "creation" had ever occurred; Voltaire similarly had no doubt that the physical universe was created by God and that God had always existed —yet both of them asked the super-ultimate "why" and regarded it as unanswerable. No doubt, some who have asked super-ultimate why-questions would, unlike Büchner and Voltaire, declare themselves ignorant of the remote history of the universe, but it is not this ignorance that they are expressing by means of the super-ultimate "why."

Those who insist that the super-ultimate why-question is meaningful do not usually deny that it very radically differs from *all* other meaningful why-questions. To mark the difference they occasionally refer to it by such labels as "mystery" or "miracle." Thus Koestenbaum remarks that "questions of this sort do not lead to answers but to a state of mind that appreciates the miracle of existence," they call attention to "the greatest of all mysteries" (*op. cit.,* pp. 54–55). Heidegger writes that the question "is incommensurable with any other" (*An Introduction to Metaphysics,* p. 4) and subsequently observes that "not only what is asked after but also the asking itself is extraordinary" (*ibid.,* p. 10).

Calling the super-ultimate why-question a "mystery" or a "miracle" or "incommensurable" or "extraordinary" does not in any way remove the difficulty: it is just one way of acknowledging that there is one. If it is granted that in all other situations a question makes sense only if an answer to it is logically possible, one wonders why this principle or criterion is not to be applied in the present case. If the defender of the meaningfulness of the super-ultimate why-question admits that in the "ordinary" sense the question is meaningless but that in some other and perhaps deeper sense it is meaningful, one would like to be told what this other and deeper sense is.

The point of the preceding paragraphs is sometimes expressed in a way that is not totally satisfactory. It is maintained that a question does not make sense unless the questioner knows what kind of answer he is looking for. However, while the fact that the questioner knows the "outline" of the answer may be a strong or even conclusive reason for supposing that the question is meaningful, the converse does not hold. One can think of examples in which a question is meaningful although the person asking it did not know what a possible answer would look like. Thus somebody might ask "What is the meaning of life?" without being able to tell us what kind of answer would be relevant and at a later time, after falling in love for the first time, he might exclaim that he now had the answer to his question—that love was the meaning of life. It would be much better to say in such a case that the question, as originally asked, was not clear than to say that it was meaningless. It is not objectionable to condemn a question as meaningless on the ground that the questioner does not know what he is looking for if in the context this is a way of saying that he has ruled out all answers *a priori*; and very probably those who express themselves in this way do not mean to point to some *contingent* incapacity on the part of the questioner but, rather, to a disability consequent upon the logical impossibility of obtaining an answer to the question. It is similar to saying that it is inconceivable that 3 plus 2 should equal 6 when we do not mean to assert a contingent fact about a certain incapacity on the part of human beings but, rather, that "3 plus 2 equals 6" is a self-contradiction.

The conclusion that the super-ultimate why-question is meaningless can also be reached by attending to what has here happened to the word "why." A little reflection shows that in the super-ultimate question "why" has lost any of its ordinary meanings without having been given a new one. Let us see how this works when the question is put in the form "Why does the universe exist?" and when the "universe" is taken to include everything that in fact exists. In *any* of its familiar senses, when we ask of anything, *x*, why it happened or why it is what it is—whether *x* is the collapse of an army, a case of lung cancer, the theft of a jewel, or the stalling of a car—we assume that there is something or some set of conditions, other than *x*, in terms of which it can be explained. We do not know what this other thing is that is suitably related to *x*, but unless it is in principle possible to go beyond *x* and find such another thing, the question does not make any sense. (This has to be slightly modified to be accurate. If we are interested in the "why" of a state of *x* at a certain time, then the answer can certainly refer to an earlier state of *x*. This does not affect the issue here discussed since, in the sense with which we are concerned, reference to an earlier state of *x* is going beyond *x*.) Now, if by "the universe" we mean the totality of things, then our *x* in "Why does the universe exist?" is so all-inclusive that it is *logically* impossible to find anything which could be suitably related to that whose explanation we appear to be seeking. "The sense of the world," wrote Wittgenstein, "must lie outside the world" (*Tractatus Logico-philosophicus*, 6.41), but by definition nothing can be outside the world. Heidegger, who avoids the formulation "Why does the universe exist?" and who instead inquires into the why of *das seiende* (the official translation of this term is "the essent," but Koestenbaum and others quite properly translate it as "things"), nevertheless makes it clear that *das seiende* here "takes in everything, and this means not only everything that is present in the broadest sense but also everything that ever was or will be." "Our question," he writes a little later, presumably without seeing the implications of this admission, "reaches out so far that we can never go further" (*An Introduction to Metaphysics,* p. 2).

For anybody who is not clearly aware of what we may call the logical grammar of "why," it is very easy to move from meaningful why-questions about particular things to the meaningless why-question about the universe. This tendency is aided by the picture that many people have of "the universe" as a kind of huge box which contains all the things "inside it." Voltaire's article "Why?" from which we quoted earlier, is a good example of such an illegitimate transition. Voltaire first asks a number of why-questions about specific phenomena, such as

> Why does one hardly ever do the tenth part good one might do? Why in half of Europe do girls pray to God in Latin, which they do not understand? Why in antiquity was there never a theological quarrel, and why were no people ever distinguished by the name of a sect?

He then gets more and more philosophical:

> Why, as we are so miserable, have we imagined that not to be is a great ill, when it is clear that it was not an ill not to be before we were born?

A little later we have what may well be a theological "why":

> Why do we exist?

Finally, as if there had been no shift in the meaning of "why," Voltaire asks:

> Why is there anything?

It should be noted that the argument we have just presented is not in any way based on an empiricist meaning criterion or on any question-begging assumptions in favor of naturalism. Anybody who uses the word "universe" in a more restricted sense, so that it is not antecedently impossible to get to an entity that might be the explanation of the universe, may be asking a meaningful question when *he* asks "Why does the universe exist?" Furthermore, even if "universe" is used in the all-inclusive sense, what we have said does not rule out the possibility that God or various divine beings are part of the universe in this sense. The point has simply been that the word "why" loses its meaning when it becomes logically impossible to go beyond what one is trying to explain. This is a matter on which there need not be any disagreement between atheists and theists or between rationalists and empiricists.

It will be well to bring together the main conclusions of this article:
(1) There is a sense in which "how" and "why" have roughly the

same meaning. In this sense science is perfectly competent to deal with the "why."

(2) There are certain senses in which "how" and "why" serve to ask distinct questions, but here too both types of questions can in principle be answered by empirical procedures.

(3) One of the cosmic "whys"—what we have called the theological "why"—is used to ask meaningful questions, at least if certain semantic problems about theological utterances are disregarded. It was pointed out, however, that this does not imply that the theological answers are true or well supported.

(4) Some apparent questions introduced by "why" are really complaints and not questions, and for this reason unanswerable.

(5) What we have called the super-ultimate "why" introduces questions that are devoid of sense, whether they are asked by ordinary people in their reflective moments or by philosophers.

❧ 17 ❧

R. M. HARE

"Nothing Matters"
Is "the Annihilation of Values" Something That Could Happen?

I

I want to start by telling you a story about something which once happened in my house in Oxford—I cannot remember now all the exact details, but will do my best to be accurate. It was about nine years ago, and we had staying with us a Swiss boy from Lausanne; he was about 18 years old and had just left school. He came of a Protestant family and was both sincerely religious and full of the best ideals. My wife and I do not read French very well, and so we had few French books in the house; but those we had we put by his bedside; they included one or two anthologies of French poetry, the works of Villon, the confessions of Rousseau and, lastly, *L'Etranger* by Camus. After our friend had been with us for about a week, and we thought we were getting to know him as a cheerful, vigourous, enthusiastic young man of a sort that anybody is glad to know, he surprised us one morning by asking for cigarettes—he had not smoked at all up till then—and retiring to his room, where he smoked them one after the other, coming down hurriedly to meals, during which he would say nothing at all. After dinner in the evening, at which he ate little, he said he would go for a walk. So he went out and spent the next three hours—as we learnt from him later—tramping round and round Port Meadow (which is an enormous, rather damp field beside the river Thames on the outskirts of Oxford). Since we were by this time rather worried about what could be on his mind, when he came back at about eleven o'clock we sat him down in an armchair and asked him what the trouble was. It appeared that he had been reading Camus's novel, and had become convinced that *nothing matters*. I do not remember the novel very well; but at the end of it, I think, the hero, who is about to be executed for a murder in which he saw no particular point even when he committed it, shouts, with intense conviction, to the priest who is trying to get him to confess and receive absolution, "Nothing matters." It was this proposition of the truth of which our friend had become convinced: *Rien, rien n'avait d'importance*.

Now this was to me in many ways an extraordinary experience. I have known a great many students at Oxford, and not only have I never known one of them affected in this way, but when I have told this story to English people they have thought that I was exaggerating, or that our Swiss friend must have been an abnormal, peculiar sort of person. Yet he was not; he was about as well-balanced a young man as you could find. There was, however, no doubt at all about the violence with which he had been affected by what he had read. And as he sat there, it occurred to me that as a moral philosopher I ought to have something to say to him that would be relevant to his situation.

Now in Oxford, moral philosophy is thought of primarily as the study of the concepts and the language that we use when we are discussing moral questions: we are concerned with such problems as "What does it mean to say that something *matters*, or *does not matter?*" We are often accused of occupying ourselves with trivial questions about words; but this sort of question is not really trivial; if it were, philosophy itself would be a trivial subject. For philosophy as we know it began when Socrates refused to answer questions about, for example, what *was* right or wrong before he had discussed the question *"What is it to be* right or wrong?"; and it does not really make any difference if this question is put in the form "What is rightness?" or "What is the meaning of the word 'right'?" or "What is its use in our language?" So, like Socrates, I thought that the correct way to start my discussion with my Swiss friend was to ask what was the meaning or function of the word "matters" in our language; what is it to be important?

He very soon agreed that when we say something matters or is important what we are doing, in saying this, is to express concern about that something. If a person is concerned about something and wishes to give expression in language to this concern, two ways of doing this are to say "This is important" or "It matters very much that so and so should happen and not so and so." Here, however, I must utter a warning lest I be misunderstood. The word "express" has been used recently as a technical term by a certain school of moral philosophers known as the Emotivists. The idea has therefore gained currency that if a philosopher says that a certain form of expression is used to *express* something, there must be something a bit shady or suspicious about that form of expression. I am not an emotivist, and I am using the word "express" as it is normally used outside philosophical circles, in a perfectly neutral sense. When I say that the words "matters" and "important" are used to express concern, I am no more committed to an emotivist view of the meaning of those words than I would be if I said "The word 'not' is used in English to express negation" or "Mathematicians use the symbol '+' to express the operation of addition."

Having secured my friend's agreement on this point, I then pointed out to him something that followed immediately from it. This is that when somebody says that something matters or does not matter, we want to know *whose* concern is being expressed or otherwise referred to. If the function of the expression "matters" is to express concern, and if concern is always *somebody's* concern, we can always ask, when it is said that something matters or does not matter, "Whose concern?" The answer to these questions is in most cases obvious from the context. In the simplest cases it is the speaker who is expressing his own concern. If we did not know what it meant in these simple cases to say that something matters, we should not be able to understand what is meant by more complicated, indirect uses of the expression. We know what it is to be concerned about something and to express this concern by saying that it matters. So we understand when anybody else says the same thing; he is then expressing his own concern. But sometimes we say things like "It matters (or doesn't matter) to *him* whether so and so happens." Here we are not expressing our own concern; we are referring indirectly to the concern of the person about whom we are speaking. In such cases, in contrast to the more simple cases, it is usual to give a clear indication of the person whose concern is being referred to. Thus we say, "It doesn't matter *to him.*" If we said "It doesn't matter," and left out the words "to him," it could be assumed in ordinary speech, in the absence of any indication to the contrary, that the speaker was expressing his *own* unconcern.

II

With these explanations made, my friend and I then returned to the remark at the end of Camus's novel, and asked whether we really understood it. "Nothing matters" is printed on the page. So somebody's unconcern for absolutely everything is presumably being expressed or referred to. But whose? As soon as we ask this question we see that there is something funny, not indeed about the remark as made by the character in the novel, in the context in which he is described as making it (though there is some-

thing funny even about that, as we shall see), but about the effect of this
remark upon my friend. If we ask whose unconcern is being expressed,
there are three people to be considered, one imaginary and two real: the
character in the novel, the writer of the novel, and its reader, my Swiss
friend. The idea that Camus was expressing his *own* unconcern about
everything can be quickly dismissed. For to produce a work of art as good
as this novel is something which cannot be done by someone who is not
most deeply concerned, not only with the form of the work, but with its
content. It is quite obvious that it mattered very much to Camus to say as
clearly and tellingly as possible what he had to say; and this argues a concern not only for the work, but for its readers.

As for the character in the novel, who thus expresses his unconcern,
a writer of a novel can put what sentiments he pleases in the mouths of
his characters—subject to the limits of verisimilitude. By the time we have
read this particular novel, it seems to us not inappropriate that the character who is the hero of it should express unconcern about absolutely
everything. In fact, it has been pretty clear right from the beginning of
the novel that he has not for a long time been deeply concerned about
anything; that is the sort of person he is. And indeed there are such people. I do not mean to say that there has ever been anybody who has literally been concerned about *nothing*. For what we are concerned about
comes out in what we choose to *do;* to be concerned about something is
to be disposed to make certain choices, certain efforts, in the attempt to
affect in some way that about which we are concerned. I do not think that
anybody has ever been *completely* unconcerned about *everything,* because
everybody is always doing something, choosing one thing rather than another; and these choices reveal what it is he thinks matters, even if he is
not able to express this in words. And the character in Camus's novel,
though throughout the book he is depicted as a person who is rather given
to unconcern, is depicted at the end of it, when he says these words, as
one who is spurred by something—it is not clear what: a sense of conviction, or revelation, or merely irritation—to seize the priest by the collar of
his cassock with such violence, while saying this to him, that they had to
be separated by the warders. There is something of a contradiction in being so violently concerned to express unconcern; if nothing *really* mattered
to him, one feels, he would have been too bored to make this rather dramatic scene.

Still, one must allow writers to portray their characters as their art
seems to require, with all their inconsistencies. But why, because an imaginary Algerian prisoner expressed unconcern for the world which he was
shortly to leave, should my friend, a young Swiss student with the world
before him, come to share the same sentiments? I therefore asked him

whether it was really true that nothing mattered to him. And of course it was not true. He was not in the position of the prisoner but in the position of most of us; he was concerned not about nothing, but about many things. His problem was not to find something to be concerned about—something that mattered—but to reduce to some sort of order those things that were matters of concern to him; to decide which mattered most; which he thought worth pursuing even at the expense of some of the others—in short, to decide what he really wanted.

III

The values of most of us come from two main sources; our own wants and our imitation of other people. If it be true that to imitate other people is, especially in the young, one of the strongest desires, these two sources of our values can be seen to have a common head. What is so difficult about growing up is the integration into one stream of these two kinds of values. In the end, if we are to be able sincerely to say that something matters for *us,* we must ourselves be concerned about it; other people's concern is not enough, however much in general we may want to be like them. Thus, to take an aesthetic example, my parents may like the music of Bach, and I may want to be like my parents; but this does not mean that I can say sincerely that I like the music of Bach. What often happens in such cases is that I *pretend* to like Bach's music; this is of course in fact *mauvaise foi*—hypocrisy; but none the less it is quite often by this means that I come in the end to like the music. Pretending to like something, if one does it in the right spirit, is one of the best ways of getting really to like it. It is in this way that nearly all of us get to like alcohol. Most developed art is so complex and remote from what people like at the first experience, that it would be altogether impossible for new generations to get to enjoy the developed art of their time, or even that of earlier generations, without at least some initial dishonesty.

Nevertheless, we also often rebel against the values of our elders. A young man may say, "My parents think it matters enormously to go to church every Sunday; but *I* can't feel at all concerned about it." Or he may say, "Most of the older generation think it a disgrace not to fight for one's country in time of war; but isn't it more of a disgrace not to make a stand against the whole murderous business by becoming a pacifist?" It is by reactions such as these that people's values get altered from generation to generation.

Now to return to my Swiss friend. I had by this time convinced him that many things did matter for him, and that the expression "Nothing

matters" in his mouth could only be (if he understood it) a piece of play-
acting. Of course he didn't actually understand it. It is very easy to assume
that all words work in the same way; to show the differences is one of the
chief ways in which philosophers can be of service to mankind. My friend
had not understood that the function of the word "matters" is to express
concern; he had thought mattering was something (some activity or process) that things did, rather like chattering; as if the sentence "My wife
matters to me" were similar in logical function to the sentence "My wife
chatters to me." If one thinks that, one may begin to wonder what this
activity is, called mattering; and one may begin to observe the world closely
(aided perhaps by the clear cold descriptions of a novel like that of
Camus) to see if one can catch anything doing something that could be
called mattering; and when we can observe nothing going on which
seems to correspond to this name, it is easy for the novelist to persuade us
that after all *nothing matters*. To which the answer is, " 'Matters' isn't
that sort of word; it isn't intended to *describe* something that things do, but
to express our concern about what they do; so of course we can't *observe*
things mattering; but that doesn't mean that they don't matter (as we can
be readily assured if, as I told my friend to do, we follow Hume's advice
and 'turn our reflexion into our own breast'[1])."

There are real struggles and perplexities about what matters most; but
alleged worries about whether anything matters at all are in most cases
best dispelled by Hume's other well-known remedy for similar doubts
about the possibility of causal reasoning—a good game of backgammon.[2]
For people who (understanding the words) say that nothing matters are,
it can safely be declared, giving but one example of that hypocrisy or
mauvaise foi which Existentialists are fond of castigating.

I am not saying that no *philosophical* problem arises for the person who
is perplexed by the peculiar logical character of the word "matters": there
is one, and it is a real problem. There are no pseudo-problems in philosophy; if anything causes philosophical perplexity, it is the philosopher's
task to find the cause of this perplexity and so remove it. My Swiss friend
was not a hypocrite. His trouble was that, through philosophical naïveté,
he took for a real moral problem what was not a moral problem at all, but
a philosophical one—a problem to be solved, not by an agonising struggle
with his soul, but by an attempt to understand what he was saying.

I am not denying, either, that there may be people who can sincerely
say that very little matters to them, or even almost nothing. We should say
that they are psychologically abnormal. But for the majority of us to be-

1. *Treatise,* III 1 i.
2. *Treatise,* I 4 vii.

come like this is a contingency so remote as to excite neither fear nor attraction; we just are not made like that. We are creatures who feel concern for things—creatures who think one course of action better than another and act accordingly. And I easily convinced my Swiss friend that he was no exception.

So then, the first thing I want to say in this talk is that you cannot annihilate values—not values as a whole. As a matter of empirical fact, a man is a valuing creature, and is likely to remain so. What may happen is that one set of values may get discarded and another set substituted; for indeed our scales of values are always changing, sometimes gradually, sometimes catastrophically. The suggestion that *nothing* matters naturally arises at times of perplexity like the present, when the claims upon our concern are so many and conflicting that we might indeed wish to be delivered from all of them at once. But this we are unable to do. The suggestion may have one of two opposite effects, one good and one bad. On the one hand, it may make us scrutinise more closely values to which we have given habitual allegiance, and decide whether we really prize them as much as we have been pretending to ourselves that we do. On the other, it may make us stop thinking seriously about our values at all, in the belief that nothing is to be preferred to anything else. The effect of this is not, as might be thought, to overthrow our values altogether (that, as I have said, is impossible); it merely introduces a shallow stagnation into our thought about values. We content ourselves with the appreciation of those things, like eating, which most people can appreciate without effort, and never learn to prize those things whose true value is apparent only to those who have fought hard to reach it. . . .

❧ 18 ❧

W. D. JOSKE

Philosophy and
the Meaning of Life

Most intelligent and educated people take it for granted that philosophers concern themselves with the meaning of life, telling us whether or not life has a meaning, and, if it has, what that meaning is. In this sense questions about the meaning of life are thought to be the direct concern of the philosopher. Philosophy is also believed to concern itself indirectly with the meaning of life, for it is thought that many purely philosophical disputes, about such topics as the existence of God, the truth or falsity of determinism, and the nature of moral judgments, can render certain attitudes to life more or less appropriate. In particular, many people are afraid of philosophy precisely because they dread being forced to the horrifying conclusion that life is meaningless, so that human activities are ultimately insignificant, absurd and inconsequential.

Although philosophy has had its great pessimists the prevailing mood among linguistic philosophers is one of qualified optimism. The mood is optimistic in so far as it is widely argued that philosophical positions cannot demonstrate the insignificance of life; but it is qualified by a caveat reminding us that if philosophical views cannot justify pessimism, they are equally incapable of providing a secure foundation for optimism. Life, it is claimed, cannot be shown to be either significant or insignificant by philosophy. In this paper I wish to argue that the contemporary attempt to establish the bland neutrality of philosophical views is not successful; philosophy is indeed dangerous stuff, and it is fitting that it should be approached with fear.

Of course the question "What is the meaning of Life?" is notoriously vague, and its utterance may be little more than an expression of bewilderment and anxiety or a shy request for help and illumination. In addition to being vague it is ambiguous. The questioner may be seeking or doubting the significance of life in general, of human life or of his own particular life. In this paper I will restrict myself to considering the significance of human life, and will not raise the questions concerning the significance of biological life or of individualistic and idiosyncratic life styles.

However, even if we restrict our concern to human life we have not removed all ambiguity. Many philosophers who have written about the meaning of human life have dealt with the significance of the course of history or the totality of human deeds, and sought to discover whether or not there was some goal which unified and made sense of all the individual and social strivings of mankind. They investigated the significance of the existence of *homo sapiens*. Yet others have wanted not the purpose of history, but a justification of the typical features of human existence, of what is fashionably called the human condition. They have asked whether or not the typical human life style can be given significance, and it is this question that I shall be concerned with. This is not to deny that there may well be connections between the two senses of the question "What is the meaning of human life?" for it may be argued that the human life style derives its significance by enabling the totality of human deeds to bring into being some valuable end. Scholastic philosophers thus see a life lived in accordance with human nature as deriving its meaning from the part which such a life will play in the fulfilment of eternal law.

If the ambiguity of our question stems from different meanings of the word "life," the vagueness grows out of the obscurity of the word "meaning." It is the vagueness of "meaning" which has enabled contemporary philosophers to dismiss pessimistic fears as unwarranted. Aware of the confusion and uncertainty that surrounds the word, they attempt to give the question "What is the meaning of life?" respectability by reading "meaning" in a precise sense borrowed from some context in which it is used with comparatively rigorous propriety. It is then discovered that the question makes no sense when interpreted in this way, so that our worries about life's meaning are shown to be pseudo-worries. The fact that somebody asks the question shows that he has not thought seriously about the meanings of the words which make it up. Nobody need live in dread of discovering that life is meaningless, just as a prudish person need not fear the vulgarity of a fraction.

The simplest and crudest example of this rejection of the legitimacy of the question is given by those philosophers who insist that only words, sentences or other conventional symbols can have meaning. Life is not a conventional symbol and is therefore neither meaningful or meaningless. Pessimism is inappropriate, and so too is optimism, but we should be cured of the desire for optimism when we realise the conceptual absurdity which it presupposes.

I do not think that any contemporary philosopher would be guilty of such a crude dismissal of the question. Intellectual fashion has changed, and in addition, we are considerably more subtle about our theories of language than we used to be. However, there is a currently popular view which, although more sophisticated and less condescending in its willingness to admit that people who puzzle about the meaning of life may be puzzling about something real, is ultimately as disparaging of the intelligence of the questioner and the seriousness of his problem as the crude argument of the past.

It is nowadays conceded that "meaning" when it occurs in the phrase "the meaning of life" is not used in the sense of conventional meaning. Things other than conventional signs may have meaning—things such as activities. A contemporary philosopher is therefore prepared to grant that a questioner may be asking if human life is meaningful in the way in which activities can be meaningful.

What is it that makes activities meaningful? It is argued that a meaningful activity is one which has significance or importance, and that the significance of an activity may either be *intrinsic,* coming from the value of a performance in itself, or *derivative,* stemming from the part which it plays in the achievement of some worthwhile end. Now it does not require much reflection to discover that people who ask about the meaning of life are not simply asking whether the standard pattern of human life is worth following for its own sake, for they are asking a question which relates both to the world and to life. They wish to know whether the world is the sort of place which justifies and gives significance to what might otherwise seem to be the drudgery of a typical human existence. In other words, they are asking whether or not the world confers derivative meaning upon life.

At this point it is easy to argue that an activity can only possess derivative value if it can be seen as bringing nearer some end which is itself worthwhile. If, as Kurt Vonnegut speculates in *The Sirens of Titan,*[1] the ultimate end of human activity is the delivery of a small piece of steel to a wrecked space ship wanting to continue a journey of no importance whatsoever, the end would be too trivial to justify the means. It is therefore appropriate to ask how the end from which life might derive its value gains its own value, and the answer which is commonly given is that even the value of this ultimate end, must, if it is to be effective, be given to it by human agents. Even if there are objective and non-natural value facts,

1. K. Vonnegut: *The Sirens of Titan* (1962).

the most worthwhile end will not satisfy an agent unless he subscribes to the value of the end, unless he commits himself to it and makes it *his* end. Even the purposes of God are useless until a man makes them his own.

It follows from this that one cannot seek the fundamental justification of human life from the nature of the world. We fail to understand the nature of meaning if we attempt to do so. The world is neutral and cannot give meaning to men. If someone wants life to be meaningful he cannot discover that meaning but must provide it himself. How we go about giving meaning to life seems to depend upon the society we accept as our own; a Frenchman might leap into the dark, an American go to a psycho-analyst, and an Englishman cease asking embarrassing questions.

The seeker after the meaning of life is thus shown to be confused, even if his confusion about the relation between fact and value is excusable. He is not however left without consolation, for if the world cannot produce meaning it cannot produce triviality. In addition, the diagnosis of his confusion directs him to the way out of his bewilderment. He must examine his own life and undertake his own commitments, even if, in order to commit himself, he may have to turn himself into a different kind of person.

An interesting variant of this contemporary optimism is proposed by Kurt Baier in his *The Meaning of Life*.[2] Baier attempts to demonstrate that any attempt to show that life is intrinsically worthless must be false, for it will depend upon criteria that are both unreasonably high and inappropriate to the evaluation of life. He argues that a worthwhile life is simply one that is above the average of the kind in respect to such things as "the balance of happiness over unhappiness, pleasure over pain, and bliss over suffering." The mere fact of its being human cannot make a life worthless, for it is necessarily true that we cannot all be below average.

It should be noted that Baier's dissolution of utter pessimism tells against the popular view that life is itself intrinsically valuable, for we cannot all be above average any more than we can all be below it. If we can assume a normal distribution of hedonic values among mankind, Baier's criteria condemn approximately half of us to lives that are not worthwhile. I would not accuse a man of undue concern if he deplored this state of affairs. However, Baier's account of the criteria of evaluation is clearly implausible. We should not describe as intrinsically worthwhile a life in which unhappiness, pain and suffering outweighed happiness, pleasure and bliss if it turned out to be the case that such a life was above the average of our kind. Again, Baier's account of evaluation has the paradoxical consequence that a man can raise himself above the average and

2. K. Baier: *The Meaning of Life* (1957) esp. pp. 25–27.

so make his life worthwhile, not only by improving himself, but also by increasing the balance of misery in the lives of other people. Clearly, the value of a life cannot be measured simply by comparing it to the average of its kind.

In spite of the plausibility of contemporary optimism I think that most of those people who question the meaning of life would not be happy with the diagnosis they have been offered. They would feel that it belittled their perplexity. All that agonising about a conceptual muddle! Perhaps their dissatisfaction is no more than pique. A neurotic who requested psychotherapy and was promptly and successfully treated with pills might well feel chagrin. He wanted to be cured, but he also wanted the adventure and discovery and struggle that are part and parcel of the sort of cure he had anticipated. He would feel less of a human being because he and his troubles were not considered worthy of analysis. Yet I do not think that the questioner's dissatisfaction is so completely unjustified. The diagnosis which he had been offered is too simple, so that the therapy suggested is not suited to the affliction from which he suffers.

I want to argue that life may be meaningless for reasons other than that it does not contribute to a worthwhile goal, so that the failure to find meaning in life can be due to the nature of the world and not simply to failure of adequate commitment by an agent. Discoveries about the world can force us to the conclusion that a committed life spent in pursuing worthwhile ends lacks meaning.

Our contemporary optimists are correct in taking the meaning of life to be analogous to the meaning of an activity, but their account of activities has been too simple. I want to argue that the significance of an activity can be challenged on grounds other than that the activity lacks intrinsic or derivative worth. To be precise, I shall claim that there are at least four elements of the meaningless, which I shall label worthlessness, pointlessness, triviality, and futility. To forestall unnecessary criticism, I must emphasise that I am not attempting to analyse the meanings of these terms as they occur in everyday discourse; I am, rather, stipulating uses of these terms which will I believe help clarify our problem.

I shall call an activity "worthless" if it lacks intrinsic merit, so that its performance needs justification by reference to some external purpose. An activity which is mere drudgery would thus be a worthless activity. Most of us would find the practice of parsing and analysing sentences worthless, and would need some extra reason for indulging in its exercise. We find ourselves bewildered by the school master in Guthrie Wilson's[3] novel,

3. Guthrie Wilson: *The Incorruptibles* (1960).

The Incorruptibles who devotes his life to parsing and analysing every sentence of *Paradise Lost.*

An activity can, of course, be meaningful even if worthless provided it is performed for some worthwhile end. An activity which is not directed towards the fulfilment of an end I shall call "pointless." A person who hits a ball against a wall without hoping to improve his health or to earn money would thus be indulging in a pointless activity. If the activity is not worthless, there is nothing absurd about it. Indeed, the very pointlessness of an activity can contribute to its pleasure, making it more truly play.

In contrast to worthless and pointless activities, an activity is trivial if, although it has a point, the purpose lacks sufficient worth to justify its performance. These are cases where the end fails to justify the means.

If lack of worth, pointlessness, and triviality were the only shortcomings which could be used to support the claim that a certain activity lacked meaning, we might indeed find it difficult to see how states of affairs outside the values of the agent could make an activity meaningless. The contemporary optimist has, however, neglected the category of the futile. This is surprising in view of the fact that so many existentialists have exalted futility into a paradigm of their beloved absurdity.

I shall call an activity "futile" if, although it has a point or needs a point in order to make it fully meaningful, the world prevents the achievement of the required end. One extreme type of futility would be an attempt to achieve some goal which is necessarily unobtainable. When we try to square the circle or solve Euler's puzzle we are guilty of such extreme futility. At the other extreme, accidental features of the world can happen to produce futility. Normally we can find uses for estimates of future population based upon current statistics and life tables, yet the task could be made futile by the unexpected disaster. In between these two extremes we find cases where the futility is neither necessary nor unpredictable. Thomas Nagel gives us the example of the man who delivers a brilliant and persuasive speech in support of a motion which has already been passed. Another case from Nagel is that of the impassioned declaration of love to a recorded announcement.[4]

The examples given above are all cases in which the world prevents the actor from realising his consciously intended end. In other cases an action may be futile although it has no consciously intended end, provided that it is the sort of activity which normally requires an end if it is to be considered worthwhile. The classical example of this sort of futility

4. T. Nagel: "The Absurd," *Journal of Philosophy* LXVIII No. 20 (Oct. 4 1971), pp. 716–727.

is that of Sisyphus—condemned to spend eternity pushing his stone to the top of a hill in order that it might roll down so that it could again be pushed to the top. A parallel case is found in the alleged practice of military prisons. A prisoner is given a small spoon and a large heap of sand. He is instructed to move the heap from one place to another, and when the task is completed to move it back again. In both these cases the punishment is peculiarly degrading because the actor is aware of the futility of the performance, and is compelled to act in accordance with desires which he cannot make his own.

We have then four different elements of the meaningful, four different ways in which we can criticise an activity for lacking in meaning.

I shall say that an activity is *fully meaningful* if it suffers from none of these defects, so that it is valuable in itself, directed towards an end which is not trivial, and is not futile. Many people who seek a short answer to the question "What is the meaning of life?" are, I suspect, looking for some simple fact or vision which will enable a human being who knows the secret to so organise his life that all of his activities will be fully meaningful. The meaning of life is thought to be the fact which makes such re-direction possible. However, few of us have such high expectations, and we are content to perform tasks that are not fully meaningful. We will endure drudgery if we can accomplish something worthwhile, and we are happy playing pointless games.

At the other extreme, I shall say that an action is *valueless* if it lacks all four elements of the meaningful, so that, from the point of view of the agent, there is no internal reason why it should be undertaken. The performance of a valueless activity is rational only if the agent is compelled to undertake it.

In between the valueless and the fully meaningful, we have those activities which are worth performing even though they fall short of the fully meaningful, such activities I shall call *valuable*.

How do our four elements of the meaningful determine the value of an action? I do not wish to explore this problem fully, but merely to make a few points. In the first place, it is clearly a tautology that if an action has worth then it is valuable. Even the performance of a futile task is justified if it is fun, and Sisyphus would have defeated the intentions of the Gods if he had happened to like rolling rocks up hills. However, it is not unnatural to hold that worth alone does not make an action completely satisfying. Many people have an incurable desire to cast a shadow across the future and affect the world so that it is forever modified by their intentions. They would find an activity which possessed only worth not nearly as rewarding as one which possessed some additional value. Those who protest concerning the irrationality of such desires should remember how we expect the activities of parents, politicians and public servants to be, in part at least, directed toward the production of benefits which will be enjoyed only by people yet unborn.

In the second place, while it is clear that triviality prevents an activity which has point gaining any additional value from that point, it is not so clear that this is the case with futility. Does an activity which is directed towards a non-trivial end, gain significance from that end if the activity is futile so that the goal can never be reached? My tentative answer is that we are each of us justified in striving for an unattainable end, provided we do not realise that the end *is* unattainable. However, when we do become aware of the futility of an activity, the goal loses its power to add meaning to the performance.

That is why the fear of futility is peculiarly haunting. We can never rid ourselves of the possibility of discovering that the world has doomed us to frustration. The altruistic project which is intended to increase the well-being of mankind may produce effects that run counter to our intentions. Subjectively, it seems that ignorance is at least near to bliss, for the ignorant man can gain satisfaction from the pursuit of the impossible, and he can derive that satisfaction only because he is ignorant.

I want now to show that certain commonly held and rationally defendable philosophical views can be seen as threatening human life with futility. I have said that I would use the phrase "human life" to refer to "the typical human life style," and this vague phrase now needs elucidation.

We expect a member of a biological species to develop in a manner which is both typical of and distinctive to the species to which it belongs. The life style of a species is the pattern of behaviour which an animal follows because it is a member of the species. When we say that cats are predatory and carnivorous we are claiming that certain types of behaviour are part of the life style of the cat. Unfortunately, if we treat human beings as a biological species, we encounter difficulties when we seek the human life style. The physiological phases of our development are clear enough, the human being will inevitably develop from childhood through adolescence to maturity and senescence, but in contrast to other animals human behaviour is pliable. There are no fixed courting rituals, no common habits of nest building, and not even any common practices of child rearing. Moreover, the variety of human practices is not akin to the variety of life styles which would be found in a collection of many different kinds of animals. *Homo sapiens* is not an assortment of species. We each of us know that if we had chosen otherwise or been brought up differently we might be enjoying or enduring a way of life very different from that which we are now living. We might have had more or fewer spouses, have left school earlier, be working in a different profession or be aiming at some achievement towards which we at present have no sympathy whatsoever.

Yet it is the very diversity of individual lives which enables us to sketch in the fundamentals of the human life style. The peculiarity of the human animal is that he can rationalise, reflect and criticise, so that he is not like a bird which cannot help but build a nest instead of a cocoon. We are creatures who pursue very general ends imposed upon us by the brute facts of physiology, and who must choose patterns of life in accordance with these general ends. The human life style thus involves critical and reflective activity, and requires the use of practical and theoretical reason directed, among other things, towards the discovery and achievement of ends which will not be seen to be futile when we learn more about our own natures and the world.

How strong is the analogy between the human life style and straightforward activities? Is there any point in assessing that life style by criteria derived from the assessment of action? It might seem not, for we do not choose either our fundamental drives or our intellectual pretensions. We find ourselves with them, rather than adopt them. A man criticising the human life style might seem as absurd as a caterpillar speculating about the value of building cocoons. We do not undertake living in the way in which we deliberately decide to do something.

Nevertheless, there is point in evaluating human life as if it were an activity. In the first place, we are not as helpless as nesting birds or metamorphosing insects. Even if the fundamentals of human life are given, there is always the choice of not living, which is why philosophers who have worried about the meaning of life have so often been obsessed by the thought of suicide. In the second place it is by no means clear that human nature cannot be changed through the interference of man. Koestler, who finds certain elements of our nature anachronistic, looks for chemicals to alter this nature.[5] Finally, even if we are only playing a game of let's pretend when we construe human life as if it were a deliberately chosen activity, many people find satisfaction in playing the game. They would find intellectual or aesthetic delight in discovering that if human life were a voluntary activity it could justifiably be chosen by a rational creature.

Having argued that there is sufficient analogy between human life and an activity to warrant our assessing them by common criteria, I want to show how philosophical views about the world may give us reason for believing in the futility of that life.

5. A. Koestler: *The Ghost in the Machine* (1968).

CASE 1. THE NAKED APE

If the speculations of some popularisers of biological theories are correct, there is good reason for holding that human life possesses the anachronistic futility exemplified by the speaker who declaims brilliantly after the motion has been passed.[6] They claim that many bewildering human activities such as our common aversion and hostility towards members of out groups, and our propensity to become violent and quarrelsome when huddled too close to our fellows, represent biological survivals which, although they once served our needs, are inappropriate to the modern environment. If our lives are necessarily an accommodation to these primitive leftovers there is a good deal of pointless struggle and torment; but the salt has yet to be rubbed into the wound.

If we see people as naked apes, we cannot but be cynical concerning the superstructure of justification associated with many of the most memorable human enterprises. Once we accept that many of our political and military endeavours are the working out of a primitive instinct of territoriality, we can no longer regard as fully meaningful the gloss of reasoning and argument which men use in an attempt to show that their undertakings are reasonable. The words of debate become mere persiflage; the talk a mere epiphenomenon of creatures ignorant of the true springs of their actions. We begin to undermine our faith in the capacity of human beings to know the truth, discover what causes what, and learn, through self-examination, about the integrity of their own motives.

CASE 2. MORAL SUBJECTIVISM

I do not believe that moral subjectivism necessarily produces a lack of accord between the world and our activities. If we were lucky, the universe and a man's attitudes could be in harmony. However, it seems that, in fact, the moral stances which people take are often of such a kind that they are futile in the face of the world. People not uncommonly sacrifice their happiness, and even their lives, for hopeless causes. Even odder than this is the fact that men who do not share a martyr's attitudes frequently admire him for standing up for truth, and the integrity of his personality.

6. See, for example, K. Lorenz: *On Aggression* (1967); Desmond Morris: *The Naked Ape: a zoologist's study of the human animal* (1967); and Robert Ardrey: *The Territorial Imperative: a personal inquiry into the animal origins of property and nations* (1967).

Good motives do not provide complete absolution, but the man who suffers in vain for his morality is commonly admired.

If we hold that ultimate value judgments can possess objective truth or falsity, and that this truth or falsity is not determined by the desires and attitudes of man, we can lend sense both to the act of self-sacrifice and to our reverence for it. The martyr is a martyr not simply for integrity but also for truth. Yet if relativistic subjectivism is true, the self-sacrifice is vain, for truth is not at stake. The martyr for the lost cause becomes, not silly, but pitiful. He accomplishes nothing and is simply a person who would have been better off had his ultimate desires been other than they were.

It will undoubtedly be said that I have missed the point of relativism. A man can have no better reason, it will be argued, for acting in a certain way than the excellent reason that what he does accords with his innermost being. It may well be the case that no reason could be better, but if this is so it is an ironic misfortune. The best reason a man has may fail to fit the nature of the universe, so that the rational act is futile.

CASE 3. ULTIMATE CONTINGENCY

I think it is still true that, among professional philosophers, the orthodox view concerning the explanation of natural phenomena is modified Humeanism. It is believed that our ability to discover the laws of nature, and to utilise them in order to both explain and cope with the world about us, depends upon the occurrence of ultimate regularities which are contingent and inexplicable. They are not inexplicable because we lack the intellectual capacity to explain, rather they are inexplicable because they have no explanation. Of course, the regularities which we encounter and recognise in our everyday dealings with the world are almost certainly not inexplicable, for we can justify our reliance upon them if we can show that they are the products of more fundamental regularities, but ultimately, it is held, the edifice of science rests upon brute and inexplicable regularities—the fundamental laws of nature.

Yet here we have a paradox. In ordinary life when we come to accept that a regularity has no explanation we regard it as a mere coincidence. If, for example, we find that all the books on a particular shelf are between 200 and 250 pages long, and we cannot discover any causal connection between their being on the shelf and their having that length, we dismiss the regularity as a mere coincidence. When we do this we acknowledge that we have no right to make any inference concerning the number of pages of the next book that might be put upon the shelf. Yet according to

Humeanism, the basic regularities which we hold to be laws of nature have no explanation. If we were to be consistent it seems that we should dismiss them as merely coincidental. Yet of course we label them laws of nature precisely because we do not wish to surrender our right to make inductive inferences concerning natural phenomena. If the world view of the Humean is correct, it seems we cannot be consistent in our attempt to construct rational models of the world. Yet reason demands consistency, so that science becomes futile; a game which cannot be won unless the rules are broken.

CASE 4. ATHEISM

The relationship between belief in God and the meaning of life is notoriously complex. Traditionally God has been used as a guarantee of objective values to provide justification for holding that certain activities are worthwhile in themselves, and also, in his capacity as designer of the universe, to justify our hope that there is point to our lives, the point deriving from the part which human existence plays in the fulfilment of the divine purpose. Unfortunately for the theist, I do not believe that anyone has been able to adequately ground objective values in the nature of God. The question which Plato asked in "The Euthyphro," "Why do the Gods value what is just?" cannot be evaded. Similarly, the attempt to give indirect value to human life by deriving that value from the purpose of God is open to the argument already mentioned, that the purposes of God are irrelevant unless we accept them as valuable in themselves and adopt them as our own. In addition, many philosophers find degrading the very idea of our lives being ordered by a superior being for the production of an end determined by the nature of that being. Thus, Kurt Baier has written, "To attribute to a human being a purpose in that sense is not neutral, let alone complimentary: it is offensive. It is degrading for a man to be regarded as merely serving a purpose."[7]

In spite of this, many people do feel that without God human life becomes meaningless. In part, the explanation of this is obvious. In most societies it is common to indoctrinate children with the belief that everything that matters derives from God. Yet what is surprising is that the view that unless there is a God life has no meaning is still proclaimed not just by children but by the sophisticated. Thus, Paul Ramsey, Professor of Religion at Princeton, is still able to claim in his *Modern Moralists* that "suicide is an inner logical consequence of *vital* atheism."[8] If Ramsey

7. K. Baier: *The Meaning of Life* (1957) p. 20.
8. P. Ramsey: *Modern Moralists* (1970) p. 26.

is correct I and many others have started people along the road to suicide. Of course he is not correct, and shows a surprising lack of awareness of the traditional arguments concerning the relation between God and morality. Nevertheless, although it does not follow from atheism that a person has no reason to continue living or to regard any activity as worthwhile, I think that atheism does open the possibility of discovering that human life possesses an extra element of absurdity. The traditional religions do claim that because the world is designed, and because a rational nature plays a part in this design, human activities cannot be futile. Although we cannot know how our desires accord with the nature of the world, the theist claims that it would be contrary to the nature of a perfect and omnipotent being to have given us desires that were not in principle capable of being acted upon and fulfilled, or to have imposed upon us life-patterns which could not be held consistent and justifiable. In this sense, I believe that although atheism does not render life meaningless, it *opens* for us the possibility of discovering that it is futile. Perhaps Descartes should not be lightly dismissed when he claims that only the theist has reason to trust his intellectual powers.

I have used these four cases to demonstrate that philosophical positions are not neutral in the conflict between pessimism and optimism, and in particular I have argued that certain widely held views lend strength to the claim that there are strong analogies between human life and a futile activity. I have not of course established that life is futile, but merely that there are *prima facie* analogies between pointless activities and many of the goal-directed undertakings that are part of the human life style. What are the practical consequences which we should draw?

In the first place, we are not justified in rejecting a philosophical view simply because if that view is true, valued life styles can no longer be uncritically accepted. We are not entitled, for example, to believe in God simply to protect our species against the charge that its activities are futile. On the other hand, views which demonstrate the fuility of the procedures of reasoning which we rely upon in order to establish that futility may be queried. Perhaps the scepticism which follows from Humeanism is a ground for holding that Humeanism must be false.

In the second place, the futility of human life does not warrant too profound a pessimism. An activity may be valuable even though not fully meaningful, and we have seen that Sisyphus would have frustrated the Gods if he could have given worth to his eternal task. We, too, can value life even if we believe that it is futile, and an active life may be fruitful in that it is productive of unexpectedly valuable consequences.[9] In addi-

9. I owe this suggestion to Professor D. A. T. Gasking.

tion, the pessimist is not by his pessimism exempted from all responsibility for his individual decisions; he is still faced with choices, and with giving meaning to the life which is the outcome of those choices.

Yet in spite of these consolations, it seems to me that it would be a matter of regret if we did discover that the basic capacities which make us human were incapable of being fulfilled. Human life would be flawed, tarnished and ultimately unreasonable; a second-best existence. When we remember that we do not choose to be human, the analogy between our lot and that of Sisyphus, the eternal prisoner, becomes grim. A philosopher, even though he enjoys living, is entitled to feel some resentment towards a world in which the goals that he must seek are forever unattainable.

EPILOGUE

BERTRAND RUSSELL

Love, Knowledge, and Pity

Three passions, simple but overwhelmingly strong, have governed my life: the longing for love, the search for knowledge, and unbearable pity for the suffering of mankind. These passions, like great winds, have blown me hither and thither, in a wayward course, over a deep ocean of anguish, reaching to the very verge of despair.

I have sought love, first, because it brings ecstasy—ecstasy so great that I would often have sacrificed all the rest of life for a few hours of this joy. I have sought it, next, because it relieves loneliness—that terrible loneliness in which one shivering consciousness looks over the rim of the world into the cold unfathomable lifeless abyss. I have sought it, finally, because in the union of love I have seen, in a mystic miniature, the prefiguring vision of the heaven that saints and poets have imagined. This is what I sought, and though it might seem too good for human life, this is what—at last—I have found.

With equal passion I have sought knowledge. I have wished to understand the hearts of men. I have wished to know why the stars shine. And I have tried to apprehend the Pythagorean power by which number holds sway above the flux. A little of this, but not much, I have achieved.

Love and knowledge, so far as they were possible, led upward toward the heavens. But always pity brought me back to earth. Echoes of cries of pain reverberate in my heart. Children in famine, victims tortured by oppressors, helpless old people a hated burden to their sons, and the whole world of loneliness, poverty, and pain make a mockery of what human life should be. I long to alleviate the evil, but I cannot, and I too suffer.

This has been my life. I have found it worth living, and would gladly live it again if the chance were offered me.

SELECTED BIBLIOGRAPHY

Althaus, Paul. "The Meaning and Purpose of History in the Christian View," *Universitas,* Vol. 7 (1965), pp. 197–204.

Ayer, A. J. "The Claims of Philosophy," in M. Natanson (ed.), *Philosophy of the Social Sciences.* New York: Random House, 1963.

Baier, Kurt. "Meaning and Morals," in Paul Kurtz (ed.), *Moral Problems in Contemporary Society.* Englewood Cliffs: Prentice-Hall, 1969.

Barnes, Hazel E. *An Existentialist Ethic.* New York: Alfred A. Knopf, Inc., 1967. Pp. 98–115 passim.

Black, Algernon. "Our Quest for Faith: Is Humanism Enough?" in P. Kurtz (ed.), *The Humanist Alternative.* Buffalo: Prometheus Books, 1973.

Blackham, H. J. "The Human Programme," in J. Huxley (ed.), *The Humanist Frame.* London: G. Allen and Unwin, 1961.

Blackham, H. J. "The Pointlessness of it All," in H. J. Blackham (ed.), *Objections to Humanism.* Phidadelphia: Lippincott, 1963.

Britton, Karl. *Philosophy and the Meaning of Life.* Cambridge: Cambridge University Press, 1969.

Brown, Delwin. "God's Reality and Life's Meaning," *Encounter,* Vol. 28, No. 3 (1968), pp. 252–62.

Brown, Delwin. "Process Philosophy and the Question of Life's Meaning," *Religious Studies,* Vol. 7 (1971), pp. 13–29.

Camus, Albert. *The Myth of Sisyphus and Other Essays,* trans. J. O. O'Brien. New York: Random House, 1955.

Clark, C. H. D. *Christianity and Bertrand Russell.* London, 1958.

de Saint Exupery, Antoine. *Wind, Sand, and Stars.* New York: Harcourt, Brace and World, 1939. Ch. 9, passim.

Dilman, Ilham. "Life and Meaning," *Philosophy,* Vol. 40 (1965), pp. 320–33.

Durant, Will (ed.) *On the Meaning of Life.* New York, 1932.

Eucken, Rudolph. *The Meaning and Value of Life.* Black, 1910.

Fackenheim, E. L. "Judaism and the Meaning of Life," *Commentary,* Vol. 39 (1965), pp. 49–55.

Flew, Antony. "Tolstoi and the Meaning of Life," *Ethics,* Vol. 73 (1963), pp. 110–18.

Frankl, Victor. *Man's Search for Meaning.* New York: Beacon, 1963.

Fromm, Erich. *Man For Himself.* New York: Holt, Rinehart, and Winston, 1947.

Greene, T. M. "Man Out of Darkness," *Atlantic Monthly*, April 1949.

Heschel, Abraham. *Man Is Not Alone*. New York: Farrar, Strauss, and Giroux, 1951.

Hocking, W. E. "Meanings of Life," *Journal of Religion*, Vol. 16 (1936).

Hollis, Christopher. "What is the Purpose of Life?" *The Listener*, Vol. 70 (1961), pp. 133–36.

James, William. "Is Life Worth Living?" in *The Will to Believe* (orig. pub. 1896). New York: Dover, 1960.

Ketcham, Charles B. *The Search for Meaningful Existence*. New York: Weybright and Talley, Inc., 1968.

Kurtz, Paul. *The Fullness of Life*. New York: Horizon Press, 1974.

Lachs, John. "To Have and To Be," *The Personalist*, Vol. 45, No 1 (1964), pp. 540–47.

Lamont, Corliss. "The Philosophy of Humanism," in C. Lamont, *The Philosophy of Humanism*. New York: Frederich Ungar, 1949. Passim.

Lewis, C. S. "De Futilitate," in *Christian Reflections*. London: Geoffrey Bles, 1967.

Maguire, J. J. "The Illogical Dr. Stace," *Catholic World*, 168 (1948), pp. 102–5.

Martineau, James. *Modern Materialism and Its Relation to Religion and Theology*. New York, 1877.

May, Rollo. *Man's Search for Meaning*. New York: W. W. Norton, 1953.

McGee, C. D. *The Recovery of Meaning—An Essay on the Good Life*. New York, 1966. Ch. 1.

Mohl, Oscar. "Man's Search for Significance in Recent Literature," *Journal of Critical Analysis*, 2 (1970), pp. 21–27.

Munitz, Milton K. *The Mystery of Existence*. New York: Appleton-Century-Crofts, 1965.

Nielsen, Kai. "Ethics Without Religion," *The Ohio University Review*, VI (1964), pp. 48–62.

Nielsen, Kai. "Religious Perplexity and Faith," *Crane Review* VIII (1965), pp. 1–17.

Nielsen, Kai. *Ethics Without God*. Buffalo: Prometheus Books, 1973.

Nielsen, Kai. "Death and the Meaning of Life," in *The Search for Absolute Values in a Changing World*, Proceedings of the Sixth International Conference on the Unity of Sciences. New York: International Cultural Foundation, 1978. Pp. 483–90.

Norton, David L. *Personal Destinies*. Princeton: Princeton University Press, 1976.

Popper, Sir Karl. "Has History Any Meaning?" in *The Open Society and Its Enemies*, 5th ed. Princeton: Princeton University Press, 1966. Vol. 2, Ch. 25.

Robinson, Richard. *An Atheist's Values*. Oxford: Oxford University Press, 1964. Pp. 54–57, 155–57.

Rosen, Stanley. *Nihilism: A Philosophical Essay*. New Haven: Yale University Press, 1969.

Russell, L. J. "The Meaning of Life," *Philosophy,* Vol. 28 (1953), pp. 30–40.

Sanders, Steven and Cheney, David R. *The Meaning of Life: Questions, Answers, and Analysis.* Englewood Cliffs: Prentice-Hall, Inc., 1980.

Sartre, Jean-Paul. "Existentialism," in *Existentialism and Human Emotions.* New York: Philosophical Library, 1948.

Schopenhauer, Arthur. "On the Sufferings of the World," trans. T. B. Saunders, in Richard Taylor (ed.), *The Will to Live: Selected Writings of Arthur Schopenhauer.* New York: Ungar, 1961.

Schopenhauer, Arthur. "On the Vanity and Suffering of Life," trans. T. B. Saunders, in Richard Taylor (ed.), *The Will to Live: Selected Writings of Arthur Schopenhauer.* New York: Ungar, 1967.

Schopenhauer, Arthur. "The Vanity of Existence," trans. T. B. Saunders, in Richard Taylor (ed.), *The Will to Live: Selected Writings of Arthur Schopenhauer.* New York: Ungar, 1967.

Stace, Walter T. "Man Against Darkness," *The Atlantic Monthly,* September, 1948.

Stern, Alfred. *The Search for Meaning: Philosophical Vistas.* Memphis: Memphis State University Press, 1971.

Tillich, Paul. *The Courage To Be.* New Haven: Yale University Press, 1952. Especially Ch. 6.

Tolstoy, Leo. *The Death of Ivan Ilych,* trans. A. Maude. New York: New American Library, 1960.

Toynbee, Arnold. "The Meaning of History for the Soul," in *Civilization on Trial.* New York: Oxford University Press, 1948. Pp. 253–63.

Troubetzkoy, Eugene. "The Meaning of Life," *The Hibbert Journal,* 16 (1918).

Warnock, G. J. *The Object of Morality.* London: Methuen, 1971.

Wittgenstein, Ludwig. *Tractatus Logico-Philosophicus,* trans. D. F. Pears and B. F. McGuinness. London: Routledge and Kegan Paul, 1961. Sections 6.4–7.

Wittgenstein, Ludwig. *Notebooks, 1914–16.* Oxford: Basil Blackwell, 1961, 1979. Pp. 72e–75e.